Lecture Notes in Networks and Systems 1466

The series "Lecture Notes in Networks and Systems" publishes the latest developments in Networks and Systems—quickly, informally and with high quality. Original research reported in proceedings and post-proceedings represents the core of LNNS.

Volumes published in LNNS embrace all aspects and subfields of, as well as new challenges in, Networks and Systems.

The series contains proceedings and edited volumes in systems and networks, spanning the areas of Cyber-Physical Systems, Autonomous Systems, Sensor Networks, Control Systems, Energy Systems, Automotive Systems, Biological Systems, Vehicular Networking and Connected Vehicles, Aerospace Systems, Automation, Manufacturing, Smart Grids, Nonlinear Systems, Power Systems, Robotics, Social Systems, Economic Systems and other. Of particular value to both the contributors and the readership are the short publication timeframe and the world-wide distribution and exposure which enable both a wide and rapid dissemination of research output.

The series covers the theory, applications, and perspectives on the state of the art and future developments relevant to systems and networks, decision making, control, complex processes and related areas, as embedded in the fields of interdisciplinary and applied sciences, engineering, computer science, physics, economics, social, and life sciences, as well as the paradigms and methodologies behind them.

Indexed by SCOPUS, EI Compendex, INSPEC, WTI Frankfurt eG, zbMATH, SCImago.

All books published in the series are submitted for consideration in Web of Science.

For proposals from Asia please contact Aninda Bose (aninda.bose@springer.com).

Amine Laghrib · Abdelghani Ghazdali
Editors

Proceedings of the Third ICMDS'24: Machine Learning, Inverse Problems and Related Fields

Springer

Editors
Amine Laghrib
Mathematics and Informatics
Sultan Moulay Slimane University
Khouribga, Morocco

Abdelghani Ghazdali
Université Sultan Moulay Slimane
Beni-Mellal, Morocco

ISSN 2367-3370 ISSN 2367-3389 (electronic)
Lecture Notes in Networks and Systems
ISBN 978-3-031-94801-5 ISBN 978-3-031-94802-2 (eBook)
https://doi.org/10.1007/978-3-031-94802-2

This Springer imprint is published by the registered company Springer Nature Switzerland AG
The registered company address is: Gewerbestrasse 11, 6330 Cham, Switzerland

Preface

Informatics plays a crucial role in advancing scientific and technological frontiers, providing powerful computational and analytical tools to address complex challenges. The rapid growth of data-driven methodologies, machine learning techniques, and computational modeling has significantly transformed modern research and applications. Among these developments, the synergy between machine learning and inverse problems has become increasingly relevant, offering innovative solutions across a wide range of scientific and industrial domains.

This special issue, *Machine Learning, Inverse Problems, and Related Fields*, focuses on the latest advances at the intersection of informatics, computational sciences, and inverse modeling. It showcases research that leverages artificial intelligence, optimization techniques, and high-performance computing to improve inverse problem-solving and predictive analytics.

The issue highlights both theoretical innovations and real-world applications, emphasizing cutting-edge methodologies that enhance computational efficiency and accuracy. The selected contributions cover diverse topics, including:

- Machine learning and deep learning techniques for solving inverse problems.
- The integration of partial differential equations (PDEs) with artificial intelligence (AI).
- Data-driven and optimization-based approaches for inverse problems.
- Computational frameworks for uncertainty quantification and predictive modeling.
- High-performance computing and algorithmic advancements in informatics.
- Applications in imaging, signal processing, bioinformatics, and environmental sciences.

By bringing together a diverse range of perspectives, this special issue underscores the transformative role of informatics in solving inverse problems and advancing machine learning methodologies. It aims to inspire further research and development in these rapidly evolving fields.

This volume reflects the growing trend of interdisciplinary research, where informatics and computational sciences serve as a bridge between diverse fields, enriching both theoretical foundations and practical applications. It brings together a collection of meticulously reviewed papers presented at the Third International Conference on Mathematical and Computational Sciences (ICMDS'24), held at ENSA Khouribga, Morocco, from October 21–22, 2024. The conference provided a dynamic platform for experts from around the world to exchange insights on recent advancements and emerging trends in machine learning, inverse problems, and computational methods, highlighting their impact across scientific and industrial domains.

This special issue reflects the ongoing evolution of informatics and computational sciences in response to modern scientific and technological challenges. It highlights how advanced computational methods, including machine learning and artificial intelligence, continue to drive innovation across diverse fields by providing powerful tools for solving complex and pressing problems. With the rapid growth of AI-driven modeling, data-driven decision-making, and high-performance computing, the contributions in this volume are expected to serve as a valuable resource for future research and applications, fostering new developments at the intersection of informatics, inverse problems, and computational sciences.

Amine Laghrib
Abdelghani Ghazdali

Contents

HMM-GMM Acoustic Modeling for Arabic Speech Recognition System

Hind Ait Mait(✉) and Noureddine Aboutabit

IPIM Laboratory, National School of Applied Sciences, Sultan Moulay Slimane University, Khouribga 25000, Morocco
hind.ait-mait@usms.ac.ma, n.aboutabit@usms.ma

Abstract. Acoustic model is a fundamental component in the construction of a speech recognition system, serving a critical function in converting raw audio signals into phonetic units that can be accurately interpreted and processed by the system. This paper presents the development of an acoustic model tailored for Arabic speech recognition system, which integrates a Hidden Markov Model (HMM) with Gaussian Mixture Models (GMM). The HMM effectively captures the temporal dynamics of speech, while the GMM computes the emission probabilities of observed acoustic features based on the hidden states. To enhance the feature representation of the model, Mel-frequency cepstral coefficients (MFCCs) are utilized alongside their first and second derivatives, capturing both the static and dynamic characteristics of the speech signal. For training and evaluation the model, the Arabic Sound Database (ASD) has been created, consisting of 34 distinct phoneme classes and includes a total of 272000 audio files. An extensive experimental evaluation was conducted to identify the optimal architectural parameters, revealing that a configuration consisting of 7 states and 16 GMM mixtures achieved the highest performance, with an accuracy of 91.47%. These findings demonstrate the efficacy of the HMM-GMM model in accurately identifying the complex acoustic patterns that are distinctive of Arabic speech.

Keywords: Arabic speech · ASR · HMM · GMM · ASD · MFCCs

1 Introduction

Speech represents one of the primary modes of human communication, functioning as a complex and dynamic channel for conveying information, emotions, and intentions. The distinctive quality of spoken language is its continuous and variable nature, which is shaped by a multitude of linguistic, phonetic, and contextual factors. This inherent complexity, coupled with the prominence of spoken interactions in everyday life, has driven a growing interest in technologies that enable machines to interpret and understand speech without manual intervention. Automatic Speech Recognition (ASR) systems are developed to

A. Laghrib and A. Ghazdali (Eds.): ICMDS 2024, LNNS 1466, pp. 1–13, 2025.
https://doi.org/10.1007/978-3-031-94802-2_1

accomplish this task, converting spoken language into text or commands for subsequent processing, thereby enabling computers to respond to human speech in a meaningful and automated manner. ASR systems operate through a series of interconnected models and processes that address the complex challenges inherent to speech recognition. One of the most important components of ASR systems is the acoustic model, which establishes a probabilistic framework that links the acoustic signal to phonetic units. This allows the system to predict which sounds correspond to specific phonemes in the language being recognized. Consequently, the accuracy and efficiency of the overall ASR system are directly influenced by the acoustic modeling process, which manages the inherent variability and noise in spoken language by accommodating differences in speech sounds across diverse speakers, dialects, and contexts. While ASR technology has advanced significantly, certain languages, such as Arabic, present unique challenges that complicate the design and effectiveness of acoustic models. Arabic in particular poses significant difficulties due to its complex phonetic structure, rich morphological system, and the prevalence of diglossia, where Modern Standard Arabic (MSA) coexists with numerous regional dialects. Additionally, Arabic speech incorporates distinctive phonemes and emphatic sounds that are not observed in many other languages, thereby further complicating the process of developing accurate acoustic representations. Furthermore, the Arabic script typically omits short vowels in the majority of texts, introducing further ambiguity in transcription and complicating the alignment between spoken and written forms. Despite the aforementioned complexities, our objective is to develop an effective acoustic model tailored for Arabic speech recognition. This paper introduces a hybrid acoustic model that combined Hidden Markov Models (HMM) and Gaussian Mixture Models (GMM) to improve recognition accuracy through optimized parameter tuning and consideration of language-specific characteristics. By utilizing Mel-Frequency Cepstral Coefficients (MFCCs) along with their first and second derivatives, the model effectively captured both the static and dynamic features of Arabic speech.

The remainder of this paper is structured as follows. Section 2 outlines some related works. Section 3 describes our proposed method and the Arabic speech database used in this study. Section 4 presents the simulation results, along with an analysis and discussion of the findings. Finally, the paper concludes with Sect. 5, which provides conclusions and outlines directions for future work.

2 Related Work

Research in speech recognition has primarily centered on English and various Asian languages. Nonetheless, in recent decades, there has been a notable rise in efforts focused on developing Arabic ASR systems, using a wide range of methods. One noteworthy contribution is presented in reference [1], which outlined an effective methodology for developing a speaker-independent continuous automatic Arabic speech recognition system. An HMM with three emission states was used as the acoustic model, with bi-grams and tri-grams incorporated into the language model in order to enhance the performance system. An

Arabic digit speech recognition system has been proposed [2], combining Artificial Neural Networks (ANN) with HMM to enhance accuracy and performance. The primary advancement lied in using an optimized neural network, specifically, a self-organizing map (SOM) to generate an optimal codebook for vector quantization, significantly boosting recognition efficiency. The paper [3] introduced a novel approach to enhance the Arabic ASR system by integrating a Support Vector Machine (SVM) with HMM. This research addressed the challenges of continuous speech recognition, with a particular focus on the accurate labeling of Arabic vowels. The primary contributions included the development of an automatic labeling method for categorizing Arabic vowels into long and short classes, and the implementation of a hybrid SVM/HMM model for the recognition of Arabic monophones and triphones. The effectiveness of the proposed method was validated through a comparative study with a traditional HMM-based system and a multi-layer perception system, which demonstrated a significant improvement in WER. Another significant contribution to the field is a hybrid model combining Learning Vector Quantization (LVQ) with HMM for Arabic phoneme recognition, as described in reference [4]. This approach demonstrated promising results, highlighting the potential of integrating diverse machine learning techniques to improve ASR accuracy. Another study introduced the TAMEEM V1.0 system [5], which leveraged a phonetically rich and balanced Modern Standard Arabic (MSA) speech corpus. This system achieved high performance across various speakers and sentence structures, underscoring the critical role of extensive, diverse training data in building robust ASR systems. In this paper [6], Arabic ASR was advanced through three key innovations: decision tree-based pronunciation modeling, hybrid acoustic modeling incorporating native models, and improved language modeling with processed text. The combined effect of these techniques led to a notable reduction in Word Error Rate. This study [7] introduced a robust ASR system utilizing Hidden Markov Models for accurately identifying Classical Arabic phonemes and allophones. To enhance system performance, the authors combined MFCC features with additional multivariate features (pitch and formants), leveraging the strengths of GMM. The proposed deep learning-based model [8] showed significant promise for accurate Arabic digit speech recognition, achieving high accuracy in both training and testing phases. This research [9] tackled the challenge of Arabic speech recognition in noisy environments by comparing two hybrid models: a GMM-HMM model implemented with CMU Sphinx and a DNN-HMM model developed with the Kaldi toolkit. Both models, using MFCCs for feature extraction, were evaluated on the Arabic Speech Corpus for Isolated Words across various noise conditions, including babble, street, and pink noise. The results indicated that the DNN-HMM model outperformed the GMM-HMM model in terms of accuracy and robustness.

3 Methodology

This section presents a description of the database used, along with a detailed overview of the features employed. It also discusses the parameters of the HMM-GMM model and the relationship between these two models.

3.1 AS Database Construction

Utilizing the Modern Standard Arabic corpus [11], which includes continuous Arabic speech, we have established the Arabic Sound Database (ASD). The primary purpose of the ASD is to train our acoustic model. As illustrated in Fig. 1, the TextGrid files associated with each audio recording have been converted into CSV format to streamline data processing. Each CSV file contains three columns: the onset time of the phoneme, the end time, and the corresponding phoneme label.

Subsequently, the timing information from the CSV file was leveraged for each WAV file in the dataset, with the objective of segmenting them into smaller WAV sub-files. These subsequently constituted isolated phoneme recordings, thereby facilitating a more focused training of the acoustic model. Following the segmentation process, the resulting audio files were organized according to their corresponding phoneme labels, placing them into distinct folders. Consequently, we have created 34 folders, each labeled with the names of the fundamental phonemes essential for training our acoustic model. This systematic organization simplifies dataset access and management throughout the training process. To mitigate any potential imbalances in the dataset during training, a data

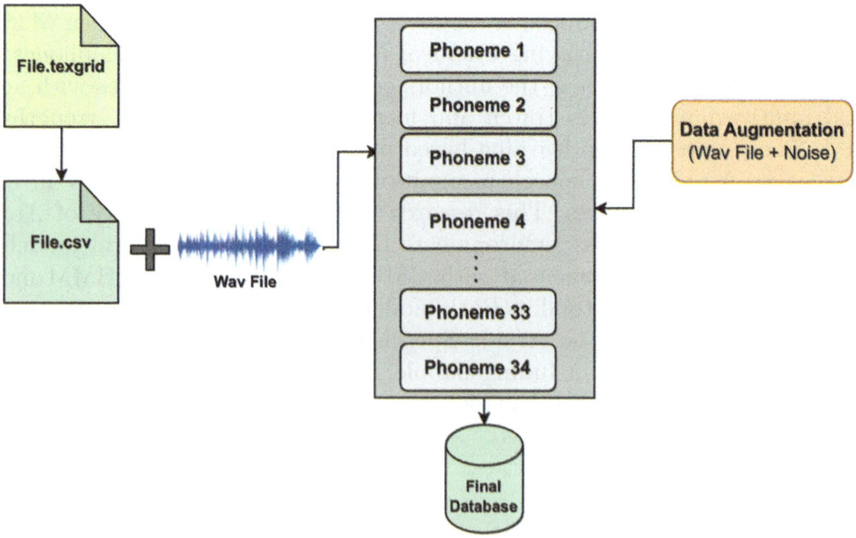

Fig. 1. Building Dataset pipeline

augmentation technique has been implemented. This technique involves enhancing data from folders with an insufficient number of files by introducing background noise into the audio signals (see Eq. 1). This augmentation process aims to achieve a more balanced distribution of data across all phoneme categories, thereby enhancing the robustness and effectiveness of our acoustic model during training.

$$S* = S + N_w \tag{1}$$

where: $S*$: the output signal, S: original signal, N_w: white noise.

Table 1 provides details about our ASD corpus, which consists of **272000** audio files, with **8000** WAV files dedicated to each phoneme. The database includes recordings of 28 consonants and 6 vowels.

Table 1. Basic phonemes.

Label	Grapheme	Utterance Number	Label	Grapheme	Utterance Number
A	ا	8000	DD	ظ	8000
B	ب	8000	AY	ع	8000
T	ت	8000	GH	غ	8000
TH	ث	8000	F	ف	8000
J	ج	8000	Q	ق	8000
HH	ح	8000	K	ك	8000
Hamza	ء	8000	L	ل	8000
KH	خ	8000	M	م	8000
D	د	8000	N	ن	8000
DH	ذ	8000	H	ه	8000
R	ر	8000	W	و	8000
Z	ز	8000	Y	ي	8000
S	س	8000	OU	اُ	8000
SH	ش	8000	I	اِ	8000
SD	ص	8000	AA	ا	8000
DE	ض	8000	OUU	و	8000
TT	ط	8000	II	ى	8000
Total				272000 utterance	

3.2 Architecture System

Our proposed methodology, illustrated in Fig. 2, outlines a systematic approach for constructing an acoustic model using a hybrid HMM-GMM framework. Initially, the database is segmented, with 80% of the data allocated for training and the remaining 20% reserved for testing, ensuring an unbiased evaluation of model performance. Features are then extracted from the training dataset, including MFCCs and their derivatives, which served as inputs for training the HMM-GMM model, chosen for its ability to accurately model both temporal dependencies and feature distributions. After training, the model is evaluated using the testing dataset to assess its generalization ability to unseen data. The same feature extraction process is employed to predict labels for the test data.

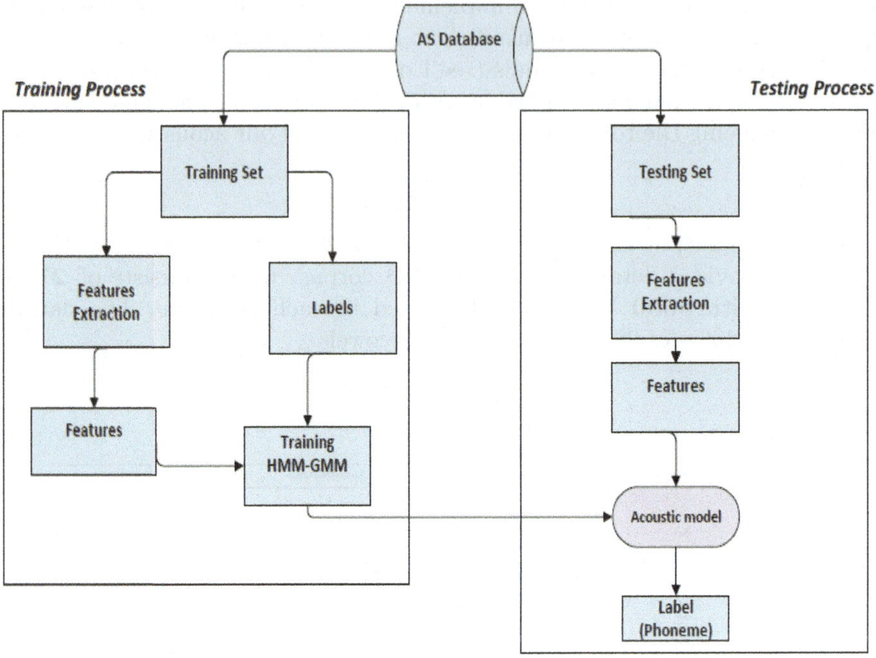

Fig. 2. Acoustic model architecture

3.2.1 Features Extraction Determining the most suitable parametric representation for acoustic signals plays a pivotal role in improving recognition accuracy. The effectiveness of this phase carries substantial implications for subsequent stages, as it profoundly impacts their functionality. In our study, we have opted for MFCCs and their derivatives as the chosen technique for feature extraction. This decision is rooted in their capability to mimic the human auditory system and their recognized potential, as evidenced by multiple studies in the field [12]. Figure 3 illustrates the complete schematic of the MFCC extraction process.

$$M(f) = 1125 \ln\left(1 + \frac{f}{700}\right) \tag{2}$$

Initially, the audio signal is divided into overlapping frames, each 30 ms in duration with a 50% overlap, and the Hamming window is applied to minimize edge effects during analysis. Following this segmentation, the Fast Fourier Transform (FFT) is employed to compute the linear spectrum of the speech signal. The resulting spectral energy is then transformed into a non-linear Mel spectrum, which aligns more closely with human auditory perception. This transformation employed a bank of Mel filters, developed from studies on the spectral properties perceived by human listeners, to produce Mel filter energies. The conversion of frequency to the Mel scale is achieved using the following equation:

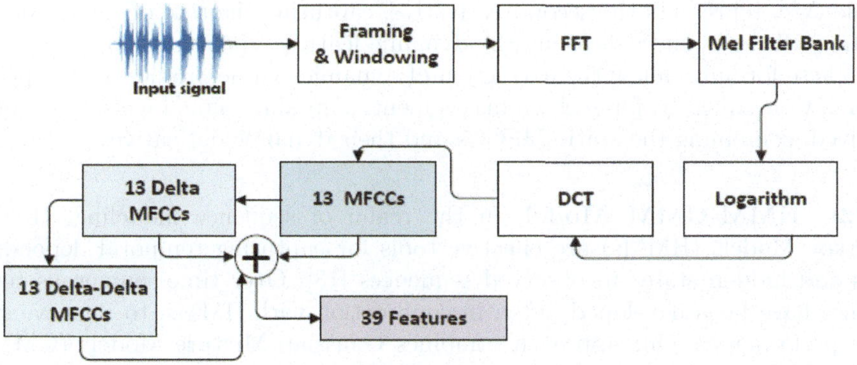

Fig. 3. MFCCs flowchart

Once the Mel filter energies are computed, a logarithmic operation is applied to compress the dynamic range and adjust the representation according to the nonlinear sensitivity of human hearing. Subsequently, these log energies are subjected to the Discrete Cosine Transform (DCT), yielding a 13-dimensional MFCC representation that captures the static features of speech. The choice of 13 MFCC coefficients is informed by empirical studies suggesting that this number effectively balances the need for sufficient detail in capturing speech characteristics while maintaining computational efficiency.

The MFCC coefficients are computed according to the formula:

$$c_i = \sum_{n=1}^{N_f} S_n \cos\left[i\left(n-\frac{1}{2}\right)\frac{\pi}{N_f}\right], \quad i = 1, 2, 3, \ldots, M \tag{3}$$

where c_i represents the i-th coefficient, N_f is the total number of filters, S_n is the log energy output of the n-th filter, and M is the desired number of MFCC coefficients. To further enhance the feature set, it is essential to include dynamic characteristics alongside the static properties of MFCCs. Therefore, the first derivative (Δ) and the second derivative (Δ^2) of the MFCCs are computed using the following formulas:

$$\Delta c_i = \frac{\sum_{j=1}^{N} j(c_{i+j} - c_{i-j})}{2\sum_{j=1}^{N} j}, \quad \text{for } i = 1, 2, \ldots, M \tag{4}$$

$$\Delta^2 c_i = \frac{\sum_{j=1}^{N} j(\Delta c_{i+j} - \Delta c_{i-j})}{2\sum_{j=1}^{N} j}, \quad \text{for } i = 1, 2, \ldots, M \tag{5}$$

where The index variable j in the summation represents the number of frames around the i-th coefficient, where c_{i+j} and c_{i-j} refer to the MFCC coefficients at positions $i + j$ and $i - j$, respectively. The notation Δc_i denotes the first derivative of the i-th MFCC coefficient, indicating its change across frames, while $\Delta^2 c_i$ represents the second derivative, capturing the rate of change of the delta coefficients and enhancing the dynamic feature representation.

These formulas allow the extraction of dynamic characteristics of the speech signal. Consequently, for each 30 ms segment, a 39-dimensional feature vector is derived, combining the static MFCCs and their dynamic derivatives.

3.2.2 HMM-GMM Model In the realm of sequence modeling, Hidden Markov Models (HMMs) are effective tools for capturing temporal dependencies and hidden states in observed sequences [13]. Over time, numerous techniques have been developed, often in conjunction with HMMs, to improve system performance. Our approach combines Gaussian Mixture Models (GMMs) with HMMs to estimate posterior probabilities. States represent underlying conditions, while the observations correspond to the features extracted from the input data, which are the observable symbols associated with each state. An N-state Hidden Markov Model is fully defined by the triplet $T = (\pi, A, B)$, where:

1. The vector $\pi = [\pi_i]$ represents the probability distribution of starting in state i.
2. The transition probability matrix $A = [a_{ij}]$ defines the likelihood of transitioning from state i to state j.
3. The emission probability matrix $B = [b_j(O_t)]$ specifies the probability distribution of observing the feature vector O_t in state j.

To enhance the expressiveness of HMMs, GMMs are incorporated for emission probabilities. Specifically, each state in the HMM is associated with a GMM, and the emission probabilities are modeled as a weighted sum of Gaussian components. This enabled the model to capture complex and diverse patterns present in the observed features. Generally, this relationship is expressed as follows:

$$b_j(O_t) = \sum_{k=1}^{K} c_{jk} \cdot N(O_t, \mu_{jk}, \Sigma_{jk}) \tag{6}$$

where K is the number of Gaussian components in the GMM, c_{jk} is the weight of the k-th Gaussian component, and $N(O_t, \mu_{jk}, \Sigma_{jk})$ is the probability density function of the k-th Gaussian component.

4 Experimental Results

The number of states and GMM mixture components are crucial hyperparameters that significantly affect both the modeling capability and recognition performance of the system. The number of states determines the temporal resolution and sequential modeling capacity of the system, with higher values allowing a more detailed representation of temporal variations in input patterns. In parallel, the number of GMM mixtures controls the complexity of acoustic modeling within each state, where increasing mixtures enhances the ability of the system to capture complex data distributions via a weighted sum of Gaussian components. These parameters function in a complementary manner: states capture temporal dynamics, whereas GMM mixtures model feature distributions within each state.

The experimental analysis depicted in Fig. 4 investigates the impact of these two critical parameters on system recognition accuracy. Our systematic evaluation encompassed three state configurations (S = 3, 5, and 7) and four GMM mixture numbers (M = 8, 12, 14, and 16), revealing several significant findings. The 3-state model, despite showing progressive improvement from 70.53% to 80.70% with increasing mixture numbers, consistently performs below the more sophisticated configurations. The 5-state architecture demonstrated substantially enhanced performance, with accuracies ranging from 89.38% to 91.05%, highlighting the importance of adequate state representation for temporal modeling. Most notably, the 7-state configuration achieved superior performance across all mixture variations, culminating in an optimal accuracy of **91.47%** with 16 mixtures.

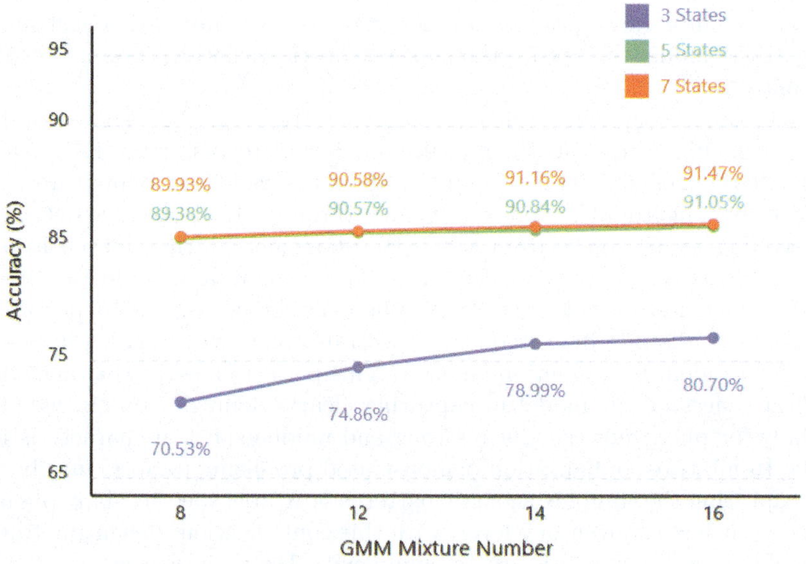

Fig. 4. Simulation Results of the Acoustic Model with Different States and GMM Mixture Numbers

This superior performance could be attributed to its enhanced capacity to model both temporal dynamics through increased state granularity and feature distribution complexity via multiple Gaussian components. Our extensive experiments revealed that while increasing the mixture number up to M = 16 yielded progressive improvements in recognition accuracy (from 89.93% at M = 8 to 91.47% at M = 16 for the 7-state model), the accuracy stabilized and maintained consistent performance when exceeding this value indicating a ceiling in recognition capability. This stability behavior is consistently observed across all state configurations, with particularly pronounced effects in the 5-state (91.05% at M = 16) and 7-state (91.47% at M = 16) models. These empirical findings provided compelling evidence that the architectural configuration comprising 7 states and 16 GMM mixtures represents the optimal parametric selection, achieving the highest accuracy of 91.47% among all tested configurations, which successfully balanced the trade-off between model complexity and recognition capability, offering optimal performance without incurring unnecessary computational overhead from additional mixture components.

In the previous analysis (Fig. 4), the performance of the acoustic model was assessed using state values of 3, 5, and 7, where the state value of **7** consistently yielded the highest average accuracy. Building upon this, the Table 2 presents a more detailed view of phoneme-level accuracy at state value 7, revealing the significant impact of the mixture number (ranging from 8 to 16) on model performance. Notably, the mixture value of 16 delivered the highest accuracy for the majority of phonemes, achieving an overall average accuracy of 91.47%. This improvement in accuracy as the number of mixtures is raised demonstrates the ability of the model to capture the subtle acoustic variations present in Arabic phonemes.

For instance, phonemes such as "KH," "HH," and "Z" show consistent improvement in accuracy as the mixture number increases, with "KH" reaching 99.69%, "HH" achieving 99.25%, and "Z" attaining 98.50% at 16 mixtures. These phonemes benefit from the ability of model to resolve finer acoustic details. The most striking result comes from the "SH" phoneme, which achieved a perfect 100% accuracy even with a mixture value of 12, maintaining this accuracy at 14 and 16 mixtures as well. The **"SH"** phoneme, being a voiceless postalveolar fricative, has a distinct high-frequency spectral profile, which makes it easier for the model to identify with minimal error. This perfect classification underscores the effectiveness of our model in capturing clearly defined acoustic signatures, especially for phonemes that have strong and unique spectral characteristics.

The 16-mixture configuration demonstrated promising performance by effectively capturing the complex acoustic patterns of Arabic speech. Many phonemes reached their highest accuracy levels with this configuration, benefiting from the ability of the model to handle greater complexity. However, for certain phonemes such as "A" and "SD," the improvement in accuracy is not as pronounced. For example, The phoneme "A" achieved an accuracy of 83.38% with 16 mixtures, which is only slightly lower than the 83.81% observed at 12 mixtures. While "SD" registered an accuracy of 79.75% at 14 mixtures, compared to 78.44% at 16 mixtures. These slight differences in performance suggest that, while the model is generally highly capable, certain phonemes exhibited less improvement with increased mixture complexity. In some cases, acoustic overlap with other phonemes or variability in pronunciation makes these sounds difficult to distin-

Table 2. Class-wise Accuracy

Phoneme	Mixture Number Values			
	8	12	14	16
A	82.88	**83.81**	82.62	83.38
B	88.38	88.94	90.19	**90.38**
T	96.12	96.62	95.94	**96.94**
TH	86.62	87.25	87.44	**87.94**
J	96.56	97.12	97.44	**97.75**
HH	98.88	99.19	99.19	**99.25**
Hamza	89.62	89.19	**90.31**	87.12
KH	99.56	99.62	99.31	**99.69**
D	79.94	84.00	82.31	**85.12**
DH	92.31	92.44	**93.94**	93.12
R	89.12	90.75	90.12	88.81
Z	97.62	97.88	97.94	**98.50**
S	92.00	93.00	92.62	**93.19**
SH	99.81	**100.00**	**100.00**	**100.00**
SD	76.50	77.75	**79.75**	78.44
DE	84.00	84.06	**88.06**	87.88
TT	87.44	88.31	**88.44**	87.62
DD	97.38	97.62	97.88	**98.50**
AY	94.88	94.44	94.06	**95.94**
GH	86.75	90.56	90.31	**94.31**
F	95.69	96.12	95.88	**97.31**
Q	97.75	98.50	98.50	**98.88**
K	95.69	96.12	96.88	**96.94**
L	90.31	91.44	**92.19**	91.56
M	87.56	88.06	**88.25**	87.88
N	83.75	82.00	85.94	**86.50**
H	92.94	95.88	**96.06**	95.12
W	90.12	90.69	92.44	**92.94**
Y	79.06	80.75	80.94	**83.56**
OU	84.69	83.38	86.19	**87.06**
I	83.62	82.25	83.06	**84.19**
AA	93.50	93.75	92.94	**94.06**
OUU	85.69	84.19	87.19	**87.31**
II	81.81	84.12	**84.50**	82.62
Average accuracy	89.93%	90.58%	91.16%	**91.47%**

guish. However, the model still demonstrated good capability, effectively managing most phonemes despite these infrequent issue.

The combination of 7 states with 16 mixtures achieved an optimal balance, effectively capturing both the temporal dynamics and acoustic variability present in Arabic phonemes. This configuration is particularly advantageous for addressing the nuanced phonetic distinctions in Arabic, which encompass a diverse range of consonants, vowels, and emphatic sounds.

5 Conclusion and Future Work

In this paper, we have successfully established a robust and efficient HMM-GMM acoustic model for Arabic speech recognition system capable of distinguishing Arabic sounds with promising accuracy. Our approach integrated MFCCs along with their first and second derivatives to enhance feature representation. Additionally, Arabic Sound Database was created to train and evaluate the model effectively. Through a series of experiments, we systematically explored various configurations of state values and Gaussian Mixture Number. Our findings indicated that the optimal configuration, provided by a state value of 7 and a GMM mixture number of 16, yielded an impressive accuracy of 91.47%. This work not only demonstrates the efficacy of our system but also contributes valuable insights into parameter optimization for Arabic speech recognition, paving the way for future advancements in the field.

As future work, we aim to integrate a deep neural network model and incorporate complementary features to further enhance the performance of the system. This direction promises to advance the capabilities of Arabic ASR systems and improve their applicability in real-world scenarios.

References

1. Abushariah, M., Ainon, R., Zainuddin, R., Elshafei, M., Khalifa, O.: Arabic speaker-independent continuous automatic speech recognition based on a phonetically rich and balanced speech corpus. Int. Arab J. Inf. Technol. **9**, 84–93 (2012)
2. Ettaouil, M., Lazaar, M., En-Naimani, Z.: A hybrid ANN/HMM models for Arabic speech recognition using optimal codebook. In: 2013 8th International Conference on Intelligent Systems: Theories and Applications (SITA), pp. 1–5 (2013)
3. Zarrouk, E., Ben Ayed, Y., Gargouri, F.: Hybrid continuous speech recognition systems by HMM, MLP and SVM: a comparative study. Int. J. Speech Technol. **17**, 223–233 (2014)
4. Nahar, K., Abu Shquier, M., Al-Khatib, W., Al-Muhtaseb, H., Elshafei, M.: Arabic phonemes recognition using hybrid LVQ/HMM model for continuous speech recognition. Int. J. Speech Technol. **19**, 495–508 (2016)
5. Abushariah, M.: TAMEEM V1. 0: speakers and text independent Arabic automatic continuous speech recognizer. Int. J. Speech Technol. **20**, 261–280 (2017)
6. Ahmed, B., Ghabayen, A.: Arabic automatic speech recognition enhancement. In: 2017 Palestinian International Conference on Information and Communication Technology (PICICT), pp. 98–102 (2017)

7. Khelifa, M., Elhadj, Y.M., Abdellah, Y., Belkasmi, M.: Constructing accurate and robust HMM/GMM models for an Arabic speech recognition system. Int. J. Speech Technol. **20**(4), 937–949 (2017). https://doi.org/10.1007/s10772-017-9456-7
8. Wazir, A., Chuah, J.: Spoken Arabic digits recognition using deep learning. In: 2019 IEEE International Conference on Automatic Control and Intelligent Systems (I2CACIS), pp. 339–344 (2019)
9. Ouisaadane, A., Safi, S.: A comparative study for Arabic speech recognition system in noisy environments. Int. J. Speech Technol. **24**(3), 761–770 (2021). https://doi.org/10.1007/s10772-021-09847-7
10. Jaber, H., Abdulbaqi, H.: Real time Arabic speech recognition based on convolution neural network. J. Inf. Optim. Sci. **42**, 1657–1663 (2021)
11. Halabi, N.: Modern standard Arabic phonetics for speech synthesis. University of Southampton (2016)
12. Singh, P., Rani, P.: An approach to extract feature using MFCC. IOSR J. Eng. **4**, 21–25 (2014)
13. Rabiner, L.: A tutorial on hidden Markov models and selected applications in speech recognition. Proc. IEEE **77**, 257–286 (1989)

SmartEdge: An Adaptive Framework for Efficient and Secure IoMT Data Processing with Real-Time Cyber Attack Detection

Anass Misbah[1(✉)], Anass Sebbar[2], and Imad Hafidi[1]

[1] National School of Applied Sciences, Computer Science and Mathematics Laboratory, Khouribga, Morocco
anassmisbah@gmail.com, i.hafidi@usms.ma

[2] International University of Rabat, ESIN, TICLab, Rabat, Morocco
anass.sebbar@uir.ac.ma

Abstract. This paper introduces SmartEdge, an advanced framework designed to enhance the processing and analysis of Internet of Medical Things (IoMT) data. SmartEdge combines systematic preprocessing, comprehensive feature selection, and an innovative model aggregation approach to minimize computational overhead, making it highly suitable for resource-constrained edge devices. Its adaptive learning capabilities enable ongoing optimization in response to new IoMT data and dynamic operational conditions.

The study demonstrates the effectiveness of feature engineering on the CICIoMT2024 dataset, achieving a substantial reduction in data dimensionality through methods such as Principal Component Analysis (PCA) and embedding techniques. These strategies lower computational loads by up to 95%, facilitating real-time data processing and enabling the deployment of machine learning models directly on edge devices. SmartEdge achieves rapid training of 10 edge models in just 55 s, maintaining an accuracy rate above 0.99 for cyber attack detection. These results highlight the framework's potential to significantly enhance IoMT security, resilience, and operational efficiency, making it a vital contribution to the development of secure and efficient IoMT ecosystems.

Keywords: IoMT security · feature selection · Machine Learning · Federated Learning · PCA

1 Introduction

The Internet of Medical Things (IoMT) is revolutionizing healthcare by enabling real-time patient data monitoring and analysis [1]. This innovation enhances patient care and operational efficiency but also introduces significant challenges [2]. Among these challenges are the demands of secure data processing and computational efficiency, particularly on edge devices with limited resources [3].

A. Laghrib and A. Ghazdali (Eds.): ICMDS 2024, LNNS 1466, pp. 14–22, 2025.
https://doi.org/10.1007/978-3-031-94802-2_2

To address these challenges, this paper presents SmartEdge, a novel framework designed to optimize IoMT security and efficiency. SmartEdge integrates systematic preprocessing, feature engineering, machine learning [4,5], and a unique model aggregation method. This holistic approach reduces computational overhead while adapting to evolving data, ensuring secure and efficient operation in resource-constrained environments.

At the core of SmartEdge is a dimensionality reduction strategy applied to the CICIoMT-2024 dataset [6]. This strategy combines feature selection [7] and feature extraction via Principal Component Analysis (PCA) [8,9]. Feature selection identifies and retains the most relevant data [10,11], while PCA condenses the dataset into a lower-dimensional form that preserves critical variance. Together, these techniques achieve a 95% reduction in data dimensionality, enabling rapid, real-time cyber attack detection on edge devices with limited computational capacity.

The effectiveness of SmartEdge is demonstrated by the rapid training of 10 edge models in just 55 s, achieving an accuracy rate exceeding 0.99 in cyber threat detection. This performance highlights SmartEdge's potential to significantly enhance IoMT cybersecurity by delivering secure, efficient, and responsive edge device functionality [12].

The remainder of this paper is organized as follows: the next section explores related work and technologies in IoMT security and dimensionality reduction. This is followed by a detailed explanation of the SmartEdge framework, including its design and implementation. The paper concludes with a discussion of findings and future research directions.

2 Related Work

2.1 Feature Engineering Techniques

SmartEdge applies advanced feature Engineering techniques to refine predictive accuracy and computational efficiency:

- **Mutual Information (MI)** [10]: MI evaluates the dependency between individual features and the target variable by quantifying the shared information. In predictive modeling, features with high MI scores reduce uncertainty about the target, making them essential for effective model building.
- **Random Forest (RF)** [7]: RF leverages an ensemble of decision trees, using the `feature_importances` attribute to rank features based on their contribution to reducing prediction uncertainty. This attribute highlights which features are most impactful, aiding in model interpretability and refinement.
- **Principal Component Analysis (PCA)** [13]: As a dimensionality reduction technique, PCA transforms features into orthogonal principal components, where high PCA loadings indicate features that capture significant data variance. This transformation reduces dimensionality while retaining crucial information for predictive performance.

- **Gradient Boosting Classifier**: This model sequentially builds weak learners to minimize prediction error, using `feature_importances` to rank features based on their influence on the loss function. This attribute provides insight into features that contribute the most to reducing error, enhancing model accuracy and interpretability.

2.2 Existing IoMT Security Frameworks

Several frameworks address the security needs of IoMT systems, often combining machine learning with privacy-enhancing approaches:

- **Explainable Machine Learning-Based Security** [14]: This framework integrates machine learning with explainable AI and federated learning to enhance IoMT security. By detecting anomalies and processing data locally, it preserves privacy, reduces data breach risks, and enables transparency in decision-making.
- **IoT-Based Healthcare Security** [15]: Focuses on real-time data analysis for IoT healthcare devices, leveraging machine learning to identify and mitigate security threats, thereby ensuring data integrity and confidentiality in healthcare IoT environments.
- **IoMT-SAF** [16]: Proposes a security assessment framework based on ontological, scenario-based analysis to identify vulnerabilities in IoMT systems. IoMT-SAF offers tailored security solutions aligned with healthcare standards, assisting stakeholders in safeguarding both medical devices and patient data.

2.3 Key Research Gaps and Challenges

The Internet of Medical Things (IoMT) faces key security challenges, including scalability, integration with healthcare standards, and adaptability to evolving technologies [17]. Existing frameworks often struggle with resource constraints, data heterogeneity, and the lack of personalized security models, leaving sensitive medical data vulnerable. Effective frameworks must focus on efficiency, interoperability, and adaptability to provide robust and reliable protection in diverse healthcare environments.

3 Main Approach

In this research, we present **SmartEdge**, an innovative framework designed to optimize the processing and analysis of Internet of Medical Things (IoMT) data using the *CICIoMT2024* dataset. The framework systematically processes data through multiple stages to enhance performance and minimize computational demands, ensuring suitability for resource-constrained environments (Fig. 1).

3.1 Data Preprocessing

SmartEdge begins with a comprehensive data preprocessing pipeline to ensure data uniformity and quality. The main steps include:

- **Noise Removal**: Elimination of inconsistencies in raw data.
- **Feature Scaling**: Standardization of feature magnitudes to ensure consistency across variables.
- **Normalization**: Harmonization of feature distributions for improved model convergence.
- **Data Integration**: Aggregation of data from diverse sources to maintain compatibility and coherence.

This preprocessing step is foundational for accurate feature selection and effective model training, particularly in the heterogeneous and dynamic IoMT environment.

3.2 Feature Selection and Dimensionality Reduction

Following preprocessing, feature selection and dimensionality reduction are applied to identify the most relevant attributes and optimize model performance. SmartEdge employs multiple techniques:

- **Random Forest (RF)**: Identifies the importance of features based on their contribution to model accuracy.
- **Principal Component Analysis (PCA)**: Reduces the dataset's dimensionality by retaining key variance components.
- **Gradient Boosting**: Enhances feature evaluation through iterative refinement.

The selection focuses on the **top five most important features**, reducing computational overhead while maintaining model effectiveness.

3.2.1 Evaluation of Feature Selection Methods Feature selection methods were evaluated for their effectiveness in terms of accuracy and computational efficiency. Table 1 presents the comparative results for the top 10 and top 5 features. The results highlight the trade-offs between accuracy and computational cost, providing insights into optimal feature selection techniques for resource-constrained IoMT devices.

3.3 Data Sampling and Dimensionality Reduction

To optimize computational efficiency, controlled data sampling and dimensionality reduction are implemented:

- **Controlled Sampling**: Data is sampled between 100 and 1,000 instances to maintain diversity while reducing computational demand.

Table 1. Comparison of Feature Selection Methods on the CICIoMT2024 Dataset

Feature Selection Method	Top 10 Accuracy	Time (Top 10)	Top 5 Accuracy	Time (Top 5)
RF Feature Importance	0.9972	23 m 9.9 s	0.9960	10 m 30 s
PCA Loadings	0.8920	66 m 61 s	0.8326	39 m 33 s
Gradient Boosting (sample of 10,000)	0.9793	12 m 39 s	0.9789	26.8 s
Mutual Information Classification	0.9979	1010 m 22.9 s	0.9976	96 m 29.2 s

- **Principal Component Analysis (PCA)**: The dataset is reduced to **two principal components**, simplifying the data structure and expediting training.

Experimental results indicate stable model accuracy across different dataset sizes, demonstrating the efficacy of sampling and PCA in enhancing computational efficiency.

3.4 Edge Model Training and Performance Evaluation

To simulate IoMT edge devices, SmartEdge trained models on balanced 1% samples of the *CICIoMT2024* dataset. Key features such as `IAT`, `Protocol Type`, and `syn_count` were prioritized. The models were trained in just 55 s with accuracies ranging from 0.9940 to 0.9950, demonstrating effective classification of varied traffic patterns.

3.5 Global Model Aggregation

To achieve robustness and generalizability, SmartEdge employs a global model aggregation strategy:

1. **Model Transfer**: Edge models are transferred to fog or cloud environments for aggregation.
2. **Performance Evaluation**: Models are re-evaluated across environments to ensure consistency.
3. **Aggregation**: Sub-models are combined into a single Random Forest model, leveraging estimator convergence to enhance robustness.

This process avoids direct data sharing, preserving privacy while improving overall performance. Experiment 4, which combined PCA and RF with the top five features, achieved the highest accuracy of 0.9984, as demonstrated in Table 1.

3.6 Results and Discussion

The evaluation results demonstrate the effectiveness of SmartEdge in achieving high accuracy with minimal computational demands. From Table 1, it is evident that Random Forest and Mutual Information Classification consistently yield the highest accuracies (0.9979) for both top 10 and top 5 features. However, Random

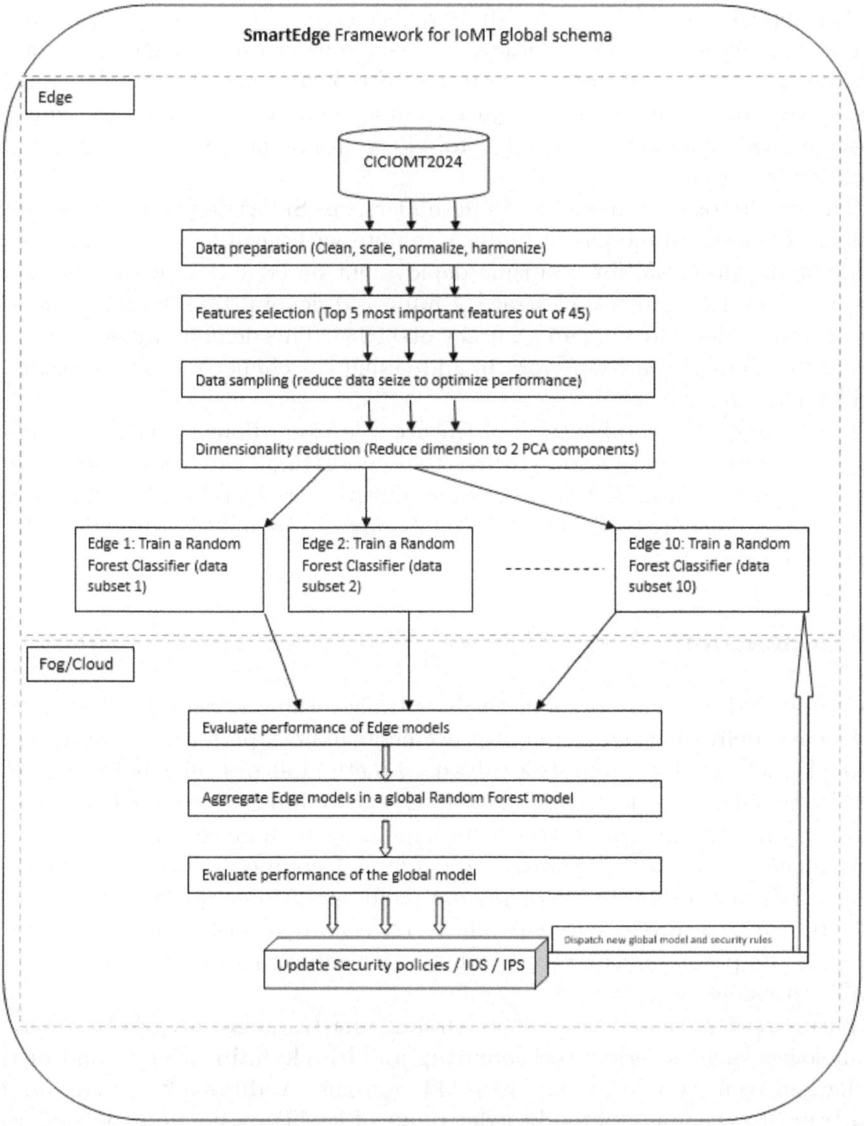

Fig. 1. SmartEdge Global Architecture Schema. The diagram illustrates the modular and hierarchical approach of SmartEdge, which integrates edge, fog, and cloud models for optimized IoMT data processing.

Forest offers a significant advantage in terms of computational time, requiring only 23 min compared to over 1000 min for Mutual Information Classification. This highlights Random Forest as the most practical feature selection technique for real-time IoMT applications.

The application of PCA successfully reduced data dimensionality while maintaining high classification accuracy. As seen during dimensionality reduction experiments, PCA achieved an accuracy of 0.9985 while drastically reducing processing time from 982 min to just 53 min. This demonstrates the suitability of PCA for resource-constrained IoMT environments, where computational efficiency is critical.

The results of edge model training underscore SmartEdge's practical applicability. Training 10 models in under a minute with accuracies exceeding 0.994 highlights its potential for real-time deployment on edge devices. Furthermore, the global model aggregation strategy improved generalizability, with the final aggregated model achieving an accuracy of 0.9984. This demonstrates the robustness of the hierarchical framework in addressing the challenges of heterogeneous IoMT environments.

In summary, the combination of feature selection, dimensionality reduction, and global model aggregation ensures that SmartEdge can deliver secure, efficient, and adaptive IoMT data processing. Future work could explore the integration of active learning to further reduce labeling efforts and improve adaptability to new data distributions.

4 Conclusion

The **SmartEdge** framework effectively optimizes the processing and analysis of IoMT data, delivering significant improvements in both performance and computational efficiency. By integrating advanced methodologies such as feature selection, Principal Component Analysis (PCA), and an innovative model aggregation strategy, SmartEdge demonstrates its capability to manage and secure IoMT environments with high accuracy and minimal computational overhead. The framework's adaptability to evolving data and operational conditions, combined with its real-time processing capabilities on resource-constrained edge devices, highlights its potential as a robust solution for enhancing cyber resilience in the IoMT ecosystem.

Future research could focus on extending SmartEdge by integrating emerging technologies such as **federated learning** and **blockchain**, which could further strengthen security and privacy in IoMT systems. Additionally, expanding the framework to accommodate a broader range of IoMT applications-beyond security and cyber attack detection-could unlock new opportunities for real-time healthcare monitoring and decision-making. Moreover, exploring the scalability of SmartEdge in complex and large-scale IoMT deployments will be essential to validate its effectiveness across diverse operational environments.

By addressing these avenues, SmartEdge has the potential to become a foundational technology in the rapidly evolving IoMT landscape, providing secure, efficient, and adaptive solutions for next-generation healthcare systems.

References

1. Dwivedi, R., Mehrotra, D., Chandra, S.: Potential of internet of medical things (IoMT) applications in building a smart healthcare system: a systematic review. J. Oral Biol. Craniofac. Res. **12**, 302–318 (2022). https://doi.org/10.1016/j.jobcr.2021.11.010
2. Kumar, A., Abhishek, K., Ghalib, M.R., Shankar, A., Cheng, X.: Intrusion detection and prevention system for an IoT environment. Digit. Commun. Netw. **8**, 540–551 (2022). https://doi.org/10.1016/j.dcan.2022.05.027
3. Gerodimos, A., Maglaras, L., Ferrag, M.A., Ayres, N., Kantzavelou, I.: IoT: communication protocols and security threats. Internet Things Cyber-Phys. Syst. **3**, 1–13 (2023). https://doi.org/10.1016/j.iotcps.2022.12.003
4. Messinis, S., Temenos, N., Protonotarios, N.E., Rallis, I., Kalogeras, D., Doulamis, N.: Enhancing internet of medical things security with artificial intelligence: a comprehensive review. Comput. Biol. Med. **170**, 108036 (2024). https://doi.org/10.1016/j.compbiomed.2024.108036
5. El Malki, I., El Hakoume, A., Laghrib, A., Hadri, A.: Improved image denoising via self-supervised weickert operator learning and plug-and-play learned primal dual. Neurocomputing **620**, 129267 (2025)
6. Dadkhah, S., Neto, E.C.P., Ferreira, R., Molokwu, R.C., Sadeghi, S., Ghorbani, A.A.: Ciciomt2024: attack vectors in healthcare devices - a multi-protocol dataset for assessing IoMT device security (2024) https://doi.org/10.20944/preprints202402.0898. Submitted to Journal of Internet of Things
7. Temghart, A.A., Marwan, M., Baslam, M.: An Improved Data Classification in Edge Cloud-Assisted IoMT: Leveraging Machine Learning and Feature Selection, pp. 156–165 (2024). https://doi.org/10.1007/978-3-031-53717-2-15
8. RM, S.P., Maddikunta, P.K.R., Koppu, S., Gadekallu, T.R., Chowdhary, C.L., Alazab, M.: An effective feature engineering for DNN using hybrid PCA-GWO for intrusion detection in IoMT architecture. Comput. Commun. **160**, 139–149 (2020). https://doi.org/10.1016/j.comcom.2020.05.048
9. Ghazdali, A., Ourdou, A., Hakim, M., Laghrib, A., Mamouni, N., Metrane, A.: Robust approach for blind separation of noisy mixtures of independent and dependent sources. Appl. Comput. Harmon. Anal. **60**, 426–445 (2022)
10. Alalhareth, M., Hong, S.-C.: An improved mutual information feature selection technique for intrusion detection systems in the internet of medical things. Sensors **23**, 4971 (2023). https://doi.org/10.3390/s23104971
11. Ourdou, A., Ghazdali, A., Laghrib, A., Metrane, A.: Blind separation of instantaneous mixtures of independent/dependent sources. Circuits Syst. Signal Process. **40**(9), 4428–4451 (2021)
12. Balla, A., Habaebi, M.H., Islam, M.R., Mubarak, S.: Applications of deep learning algorithms for supervisory control and data acquisition intrusion detection system. Clean. Eng. Technol. **9**, 100532 (2022). https://doi.org/10.1016/j.clet.2022.100532
13. Rupapara, V., Rustam, F., Ishaq, A., Lee, E., Ashraf, I.: Chi-square and PCA based feature selection for diabetes detection with ensemble classifier. Intell. Autom. Soft Comput. **36**(2), 1931–1949 (2023). https://doi.org/10.32604/iasc.2023.028257
14. Si-ahmed, A., Al-Garadi, M.A., Boustia, N.: Explainable machine learning-based security and privacy protection framework for internet of medical things systems (2024)

15. Pirbhulal, S., Pombo, N., Felizardo, V., Garcia, N., Sodhro, A.H., Mukhopadhyay, S.C.: Towards machine learning enabled security framework for IoT-based healthcare. In: 2019 13th International Conference on Sensing Technology (ICST), pp. 1–6. IEEE (2019). https://doi.org/10.1109/ICST46873.2019.9047745
16. Alsubaei, F., Abuhussein, A., Shandilya, V., Shiva, S.: IoMT-SAF: internet of medical things security assessment framework. Internet Things **8**, 100123 (2019). https://doi.org/10.1016/j.iot.2019.100123
17. Ghubaish, A., Salman, T., Zolanvari, M., Unal, D., Al-Ali, A., Jain, R.: Recent advances in the internet-of-medical-things (IoMT) systems security. IEEE Internet Things J. **8**, 8707–8718 (2021). https://doi.org/10.1109/JIOT.2020.3045653

Machine Learning Prediction of Long Jump Performance Based on Biomechanical Factors

Marouane Baadi[1(✉)], Lekbir Afraites[1], and Soufiane Lyaqini[2]

[1] EMI, FST, Sultan Moulay Slimane University, Beni Mellal, Morocco
{marouane.baadi,l.afraites}@usms.ma

[2] Ecole Nationale des Sciences Appliquées, Sultan Moulay Slimane University, Khouribgha, Morocco

Abstract. This study investigates the use of machine learning to predict long jump performance based on biomechanical features. Data from nineteen biomechanical variables collected from elite long jumpers were analyzed, with machine learning models including Random Forest, CatBoost, Gradient Boosting, and XGBoost used for performance prediction. Feature importance analysis, conducted using the Random Forest model, highlighted key predictors such as gender, horizontal velocities during critical strides leading up to take-off, and vertical velocity at take-off. Among the models tested, XGBoost demonstrated the highest accuracy, achieving an R^2 of 0.9454, MAE of 0.1552, and RMSE of 0.1791. These findings emphasize the pivotal role of biomechanical factors in predicting performance and offer data-driven insights for enhancing long jump outcomes for athletes and coaches.

Keywords: Long jump performance · Biomechanics · Machine learning · Feature importance · Predictive modeling · Athletic performance optimization

1 Introduction

The long jump is a technically demanding athletic event where performance is influenced by a multitude of biomechanical factors. Traditionally, coaching in this domain has relied on observational analysis and experiential knowledge. However, advancements in machine learning now offer a quantitative framework to optimize performance [1]. By leveraging biomechanical data, machine learning models can quantify the contributions of various physical variables, enabling precise performance predictions [2,3].

This study presents a machine learning-based approach to predicting long jump performance using nineteen biomechanical variables, such as horizontal and vertical velocities, flight times, and body mechanics. Four machine learning

L. Afraites and S. Lyaqini—Equal contribution.

A. Laghrib and A. Ghazdali (Eds.): ICMDS 2024, LNNS 1466, pp. 23–35, 2025.
https://doi.org/10.1007/978-3-031-94802-2_3

models-Random Forest, CatBoost, Gradient Boosting, and XGBoost-were evaluated, with XGBoost demonstrating the highest predictive accuracy [4,5]. The findings of this research not only underscore the importance of critical biomechanical factors but also provide actionable insights for athletes and coaches aiming to enhance performance outcomes.

The long jump consists of several distinct phases, beginning with the approach run. Elite jumpers typically execute 18 to 22 controlled strides to achieve optimal velocity and stability before takeoff. The takeoff phase, initiated at the "touchdown" point when the takeoff foot contacts the ground, is critical for converting horizontal velocity into vertical impulse. Achieving an ideal takeoff angle, generally between 21° and 27°, is essential to balance horizontal speed with vertical lift, maximizing jump distance (Linthorne). High horizontal velocities are prioritized, as research by Lees et al. (1994), Seyfarth, Blickhan, and Van Leeuwen (2000), Alexander (1990) [6], and Panoutsakopoulos (2010) [7] has demonstrated a strong correlation between takeoff speed and jump distance. Recent studies, such as that by De Lazzari, Mascia, and Vannozzi (2023), have further advanced performance analysis by using machine learning to identify key biomechanical factors, emphasizing the transformative role of technology in athletic performance analytics [8].

This machine learning approach demonstrates the potential for data-driven optimization of long jump training, revolutionizing traditional coaching practices with predictive analytics derived from biomechanical insights.

The structure of this paper is as follows. In Sect. 2, we describe the dataset, detailing biomechanical metrics relevant to long jump analysis, along with preprocessing steps, model training (focusing on Random Forest and XGBoost), and evaluation metrics (R^2, MAE, RMSE) used to assess predictive accuracy. Section 3 elaborates on the materials and methods employed. Finally, Sect. 4 presents the conclusions and implications of this study.

2 Material and Methods

This section provides a detailed description of the dataset, emphasizing biomechanical metrics critical for analyzing long jump performance. Preprocessing steps, such as label encoding, handling missing values, and data splitting for training and testing, are outlined. The machine learning models employed-Random Forest and XGBoost-are discussed, along with the evaluation metrics (R^2, MAE, RMSE) used to assess model accuracy. Lastly, the target feature, jump distance, is highlighted, and the insights derived from these analyses are discussed in the context of enhancing athletic performance.

2.1 Dataset Description

The dataset [9] offers a comprehensive analysis of biomechanical performance metrics in long jump athletes. It includes critical variables such as step lengths, contact and flight times, horizontal and vertical velocities, joint angular velocities, and angular dynamics of the trunk and body. Additional features, such as trunk and body inclination angles, knee angular velocity, and swing leg mechanics, provide a detailed view of the kinematic and kinetic factors influencing long jump performance.

This dataset also includes gender and official jump distance measurements, enabling analysis of how these biomechanical variables impact outcomes across athletes of different profiles. Likely compiled from a mix of training footage and publicly available media clips of elite athletes, the dataset provides valuable insights for optimizing long jump techniques and enhancing training methodologies (Table 1).

Table 1. Description of the Dataset Columns

Column Name	Description
Gender	Athlete's gender (Male/Female)
Takeoff Loss	Energy or velocity loss at takeoff
3rd Last Step Length	Length of the third-to-last step before takeoff
3rd Last Step Horizontal Velocity BTO (m/s)	Horizontal velocity during the third-to-last step (before takeoff)
2nd Last Step Horizontal Velocity BTO (m/s)	Horizontal velocity during the second-to-last step (before takeoff)
Last Step Horizontal Velocity BTO (m/s)	Horizontal velocity during the last step (before takeoff)
Horizontal Velocity at Takeoff (m/s)	Horizontal velocity at the moment of takeoff
Vertical Velocity at Takeoff (m/s)	Vertical velocity at the moment of takeoff
3rd Last Step Contact Time (s)	Ground contact time for the third-to-last step
3rd Last Step Flight Time (s)	Flight time for the third-to-last step
2nd Last Step Contact Time (s)	Ground contact time for the second-to-last step
Trunk Angle at Takeoff (°)	Trunk angle at the moment of takeoff
Body Inclination Angle at Touchdown (°)	Body inclination angle at touchdown
Mean Knee Angular Velocity at TD (°/s)	Average knee angular velocity at touchdown
Thigh Angular Velocity of Swing Leg at TO (°/s)	Angular velocity of the swing leg's thigh at takeoff
Official Distance (m)	Measured jump distance in meters

2.2 Dataset Preprocessing

Effective preprocessing was applied to prepare the dataset for machine learning, ensuring compatibility and optimizing performance.

- **Label Encoding:** The categorical variable "Gender" was transformed into numerical values using `LabelEncoder`. This conversion is essential for machine learning algorithms that require numerical input.
- **Handling Missing Values:** Missing values in both the feature matrix (X) and the target variable (y) were imputed using the mean value of the respective column. This simple imputation method ensures data completeness and avoids errors during model training.

2.2.1 Splitting Data Into Training and Testing Sets

To evaluate model performance, the dataset was split into training and testing subsets.

- **Data Splitting:** The `train_test_split` function was used to partition the dataset, allocating 75% of the data for training and 25% for testing.
- **Objective:**
  - The training set was used to train the models and learn relationships between features and the target variable.
  - The testing set was reserved for evaluating how well the models generalize to unseen data.
- **Parameters:**
  - `test_size=0.25`: Allocates 25% of the dataset for testing purposes.
  - `random_state=42`: Ensures reproducibility by setting a fixed random seed for data shuffling.

2.2.2 Training and Predicting Models

This stage involves training various machine learning models and using them to make predictions on unseen data.

- **Model Training:** Each regressor-Random Forest, XGBoost, CatBoost, and Gradient Boosting-was instantiated with specific hyperparameters tailored to optimize performance. The models were trained on the training dataset (`train_X` and `train_y`), learning the relationships between the input features (`train_X`) and the target variable (`train_y`).
- **Prediction on Test Data:** After training, the models were evaluated on the test dataset (`test_X`) to generate predictions (`pred_y`). This process assesses how well the models generalize to new, unseen data.

2.2.3 Model Evaluation

To assess the performance of each model, three key metrics were used:

- **R^2 Score** (R^2`_score`): Indicates the proportion of variance in the target variable that is predictable from the input features. A value closer to 1 represents a better fit, with 1 indicating perfect alignment between predictions and actual values.
- **Mean Absolute Error (MAE):** Represents the average magnitude of errors between predicted and actual values, calculated without considering the direction of errors. MAE provides an intuitive measure of the model's prediction accuracy.
- **Root Mean Squared Error (RMSE):** Similar to MAE, RMSE measures the average magnitude of errors but gives greater weight to larger errors by squaring them before averaging. This makes RMSE particularly sensitive to significant deviations and useful for penalizing models with large prediction errors.

2.3 Target Feature

The target feature in this machine learning experiment is the long jump distance, measured in meters, denoted as y. The dataset used, contains multiple features such as gender, athlete characteristics, and jump-related parameters. The model's goal is to predict the long jump distance based on these input features.

2.4 Evaluation Metrics

The performance of the machine learning models was evaluated using several key metrics. These metrics help to assess the accuracy and overall fit of the models to the data. Below is a description of each metric used, along with the performance results of the XGBoost model, which showed the best predictive accuracy.

2.4.1 Mean Absolute Error (MAE)

[10] The Mean Absolute Error (MAE) measures the average magnitude of the errors in a set of predictions, without considering their direction. It is calculated as:

$$\text{MAE} = \frac{1}{n}\sum_{i=1}^{n} |y_i - \hat{y}_i|$$

A lower MAE value indicates that the model's predictions are closer to the actual values.

2.4.2 Root Mean Squared Error (RMSE)

[11] The Root Mean Squared Error (RMSE) is similar to MAE but provides a higher penalty for larger errors. It is calculated as:

$$\text{RMSE} = \sqrt{\frac{1}{n}\sum_{i=1}^{n} (y_i - \hat{y}_i)^2}$$

A lower RMSE indicates that the model is performing well, especially in cases where larger errors are undesirable.

2.4.3 Coefficient of Determination ($\mathbf{R^2}$)

[12] The Coefficient of Determination (R^2) measures the proportion of variance in the dependent variable that is predictable from the independent variables. It is calculated as:

$$R^2 = 1 - \frac{\sum_{i=1}^{n} (y_i - \hat{y}_i)^2}{\sum_{i=1}^{n} (y_i - \bar{y})^2}$$

An R^2 value closer to 1 indicates a better fit of the model to the data.

3 Results and Discussion

This section presents a comparative analysis of four machine learning models-Random Forest [13], XGBoost [14], CatBoost [15], and Gradient Boosting [16]-using three key metrics: R^2, MAE, and RMSE. The primary objective is to evaluate the models' ability to predict long jump distances accurately. Among the models, **XGBoost** consistently outperforms the others, achieving the highest R^2 (**0.6917**), the lowest MAE (**1.1538**), and the lowest RMSE (**1.4633**).

Feature importance analysis further highlights the critical role of horizontal velocities in approach steps across all models. However, XGBoost shows a bias toward "Gender," in contrast to the more balanced feature distributions observed in Random Forest and Gradient Boosting [17,18]. This study provides a unique contribution by identifying and quantifying the influence of key biomechanical features, such as horizontal velocities during the approach phase, in predicting long jump performance. Furthermore, it emphasizes the potential pitfalls of biased feature weighting in machine learning models, offering insights for more equitable and generalizable predictive frameworks in sports analytics.

3.1 Performance Metrics Comparison: R^2, MAE, and RMSE Across Models

The performance of the machine learning models was evaluated based on the following metrics:

1. R^2 (Coefficient of Determination)

- **XGBoost** achieves the highest R^2 score of **0.6917** (Fig. 7), indicating superior ability to explain the variance in the test data.
- **Gradient Boosting** ranks second with an R^2 of **0.6837** (Fig. 5), followed by **CatBoost** with **0.6615** (Fig. 1).
- **Random Forest** scores the lowest R^2 at **0.6499** (Fig. 3).

A higher R^2 value signifies better model performance, with XGBoost emerging as the top model.

2. MAE (Mean Absolute Error)

- **XGBoost** achieves the lowest MAE of **1.1538**, indicating its predictions are closest to the actual values.
- **CatBoost** and **Gradient Boosting** have comparable MAEs of **1.1684** and **1.1755**, respectively.
- **Random Forest** performs the worst with an MAE of **1.1939**.

Lower MAE values signify better prediction accuracy, with XGBoost leading the pack.

3. RMSE (Root Mean Squared Error)

- **XGBoost** delivers the lowest RMSE of **1.4633**, followed by **Gradient Boosting** (**1.4854**) and **CatBoost** (**1.5346**).

- **Random Forest** has the highest RMSE at **1.5545**.

Lower RMSE values indicate fewer large errors in predictions, further reinforcing XGBoost's superiority.

3.2 Analyzing Feature Importance in Long Jump Performance Prediction Models

The analysis in Table 2 shows that XGBoost explains 94.54% of the variance in the target variable, with a mean absolute error of 0.1552 and a root mean squared error of 0.1791. These metrics underscore its exceptional accuracy.

1. Velocity Features
Horizontal velocities, particularly those for the "3rd last step," "2nd last step," and "last step," were consistently ranked as critical features across models. Vertical velocity at takeoff was moderately important, highlighting the significance of approach speed and takeoff dynamics in predicting long jump performance.

2. Balanced vs. Skewed Feature Importance Distributions

- **XGBoost:** The model (Fig. 8) heavily prioritizes "Gender," which may introduce bias. It relies less on athletic metrics, leading to potential limitations in generalizing predictions.
- **Random Forest and Gradient Boosting:** These models (Figs. 4 and 6) display a more balanced distribution of feature importance, effectively combining "Gender" with velocity metrics for comprehensive prediction accuracy.
- **CatBoost:** Although "Gender" ranks highly, CatBoost balances this with horizontal velocity metrics, ensuring a nuanced approach to evaluating athletic performance (Fig. 2).

Table 2. Comparison of Model Performance on Test Set.

Metric	RandomForest	XGBoost	CatBoost	GradientBoosting
R^2 (Test)	0.6499	0.6917	0.6615	0.6837
MAE (Test)	1.1939	1.1538	1.1684	1.1755
RMSE (Test)	1.5545	1.4633	1.5346	1.4854

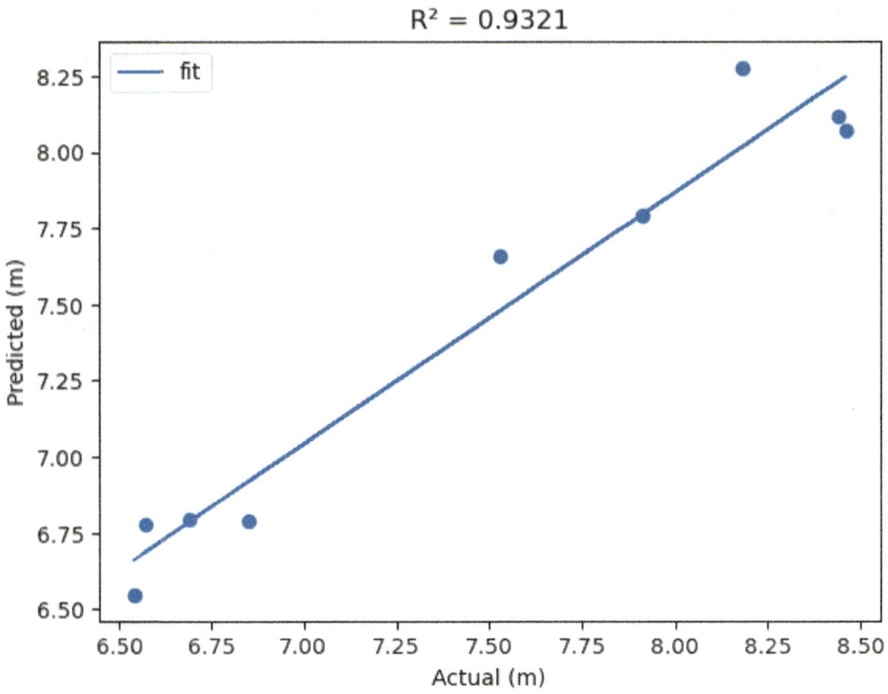

Fig. 1. Actual vs. Predicted Long Jump Distances using CatBoost Model.

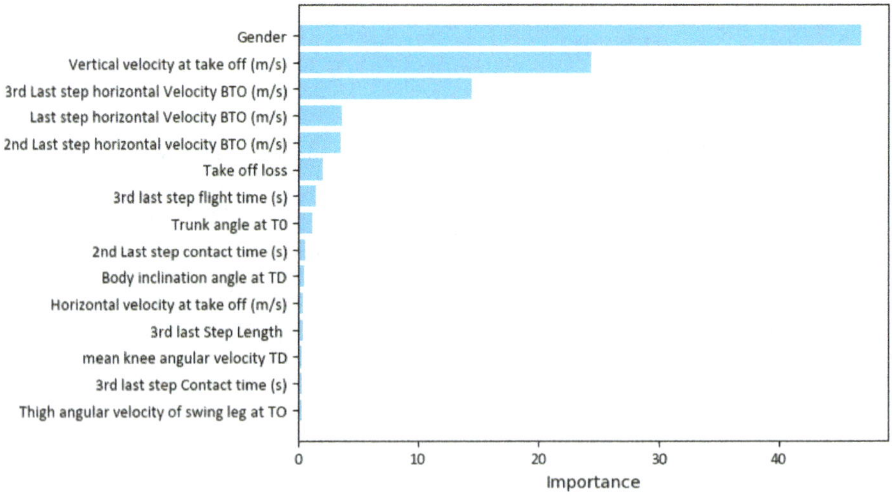

Fig. 2. Feature Importance for Long Jump Distance Prediction using CatBoost Model.

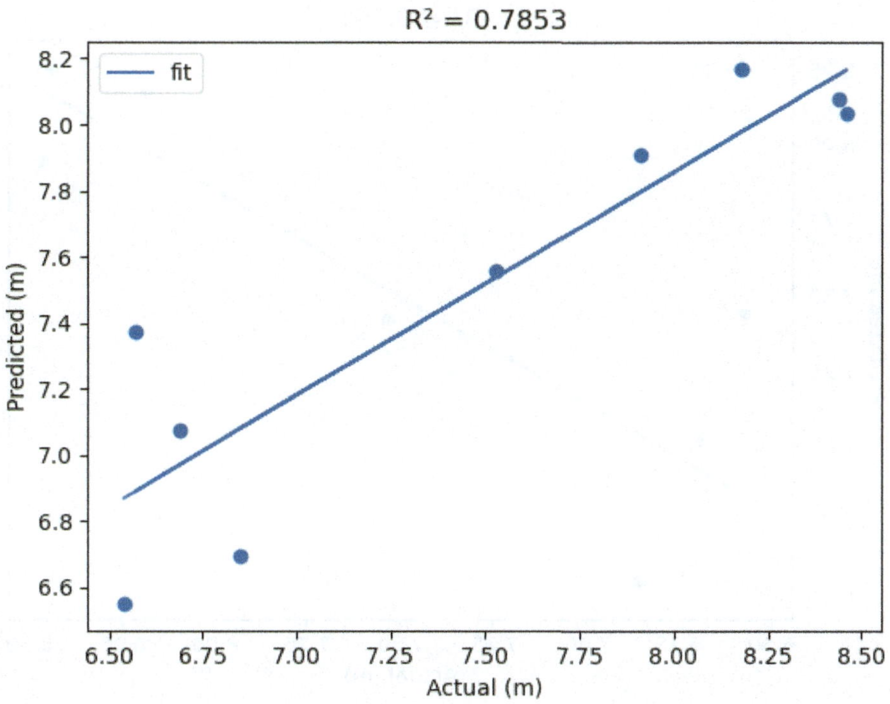

Fig. 3. Actual vs. Predicted Long Jump Distances using Random Forest Model.

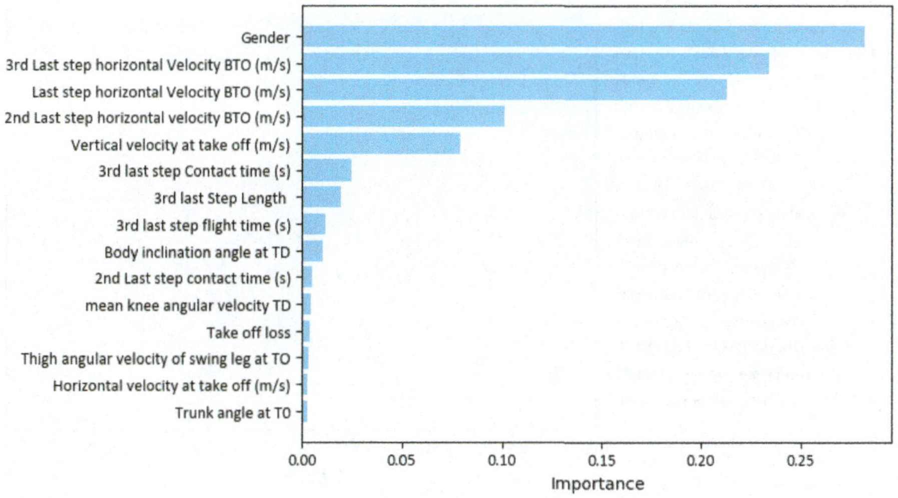

Fig. 4. Feature Importance for Long Jump Distance Prediction using Random Forest Model.

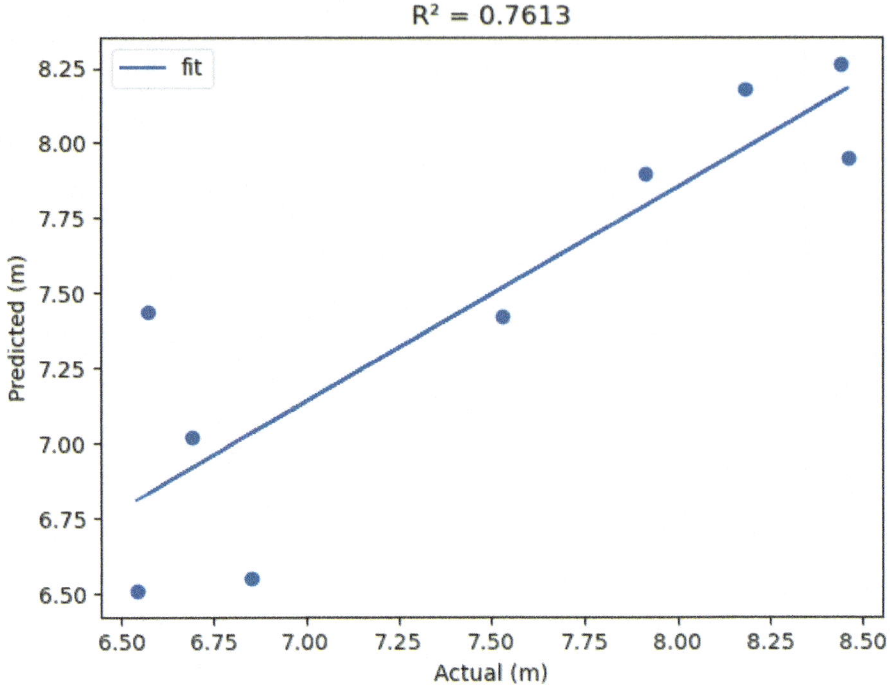

Fig. 5. Actual vs. Predicted Long Jump Distances using Gradient Boosting Model.

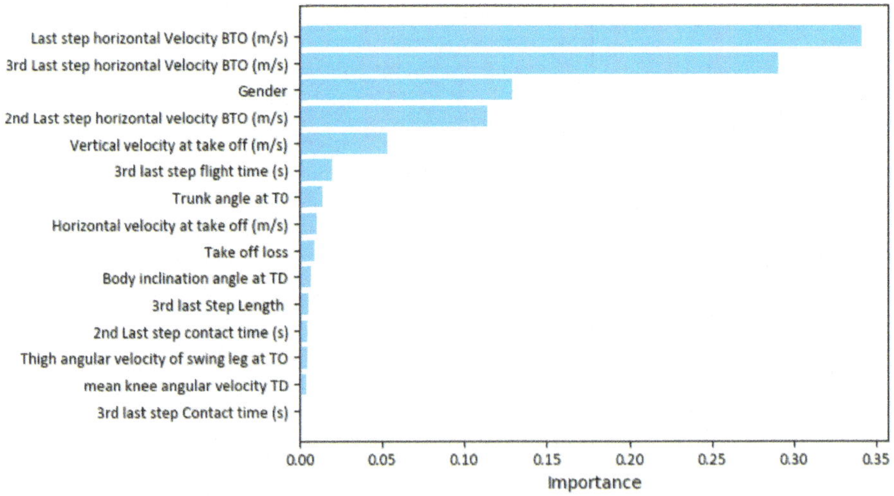

Fig. 6. Feature Importance for Long Jump Distance Prediction using Gradient Boosting Model.

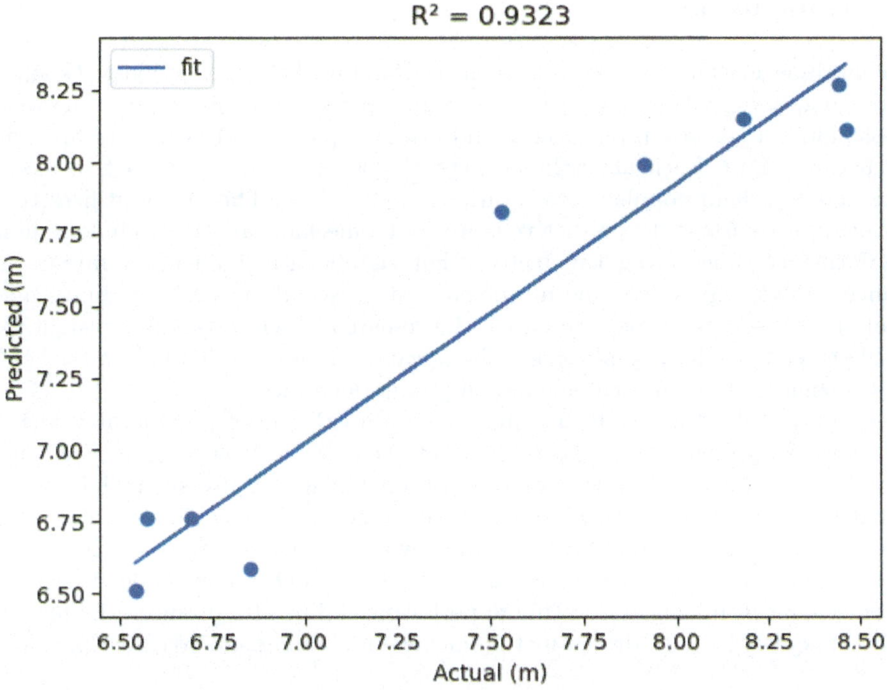

Fig. 7. Actual vs. Predicted Long Jump Distances using XGBoost Model.

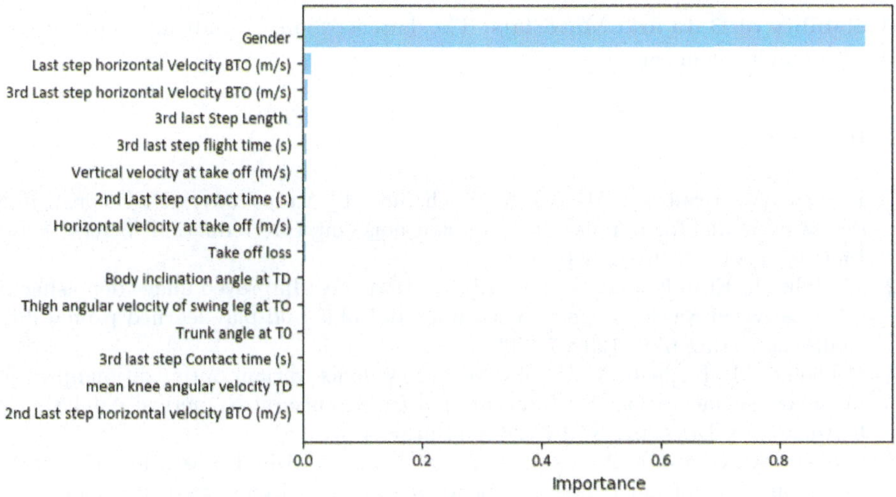

Fig. 8. Feature Importance for Long Jump Distance Prediction using XGBoost Model.

4 Conclusion

In machine learning, algorithms such as Random Forest, CatBoost, Gradient Boosting, and XGBoost are widely recognized for their robust performance in tasks like classification, regression, and ranking, particularly when dealing with structured data. Each algorithm brings unique strengths to predictive modeling, making them popular across various applications. This study utilizes these algorithms to assess the predictive power of biomechanical features in long jump performance, uncovering key factors that significantly influence jumping distance. By identifying specific movement patterns and physical attributes associated with successful performances, the research provides valuable insights for athletes and coaches, emphasizing the advantages of data-driven approaches in optimizing training and enhancing long jump outcomes.

Additionally, this study highlights the critical role of horizontal velocities during the approach phase as a dominant predictor of long jump performance, providing a quantified understanding of their impact. It also sheds light on the challenges associated with biased feature weighting in machine learning models, urging the adoption of balanced frameworks to ensure fair and generalizable insights in sports analytics. These findings not only deepen our understanding of biomechanical influences on athletic performance but also promote the integration of equitable, data-driven methodologies in training and coaching practices.

Statements and Declarations

Conflict of Interest. The authors declare that they have no conflict of interest.

Availability of Data and Materials. The data used in this study are available from https://olympics.com/en/.

References

1. Laghrib, A., Afraites, L., Hadri, A., Nachaoui, M.: A non-convex PDE-constrained denoising model for impulse and gaussian noise mixture reduction. Inverse Probl. Imaging **17**(1), 23–67 (2023)
2. El Malki, I., El Hakoume, A., Laghrib, A., Hadri, A.: Improved image denoising via self-supervised weickert operator learning and plug-and-play learned primal dual. Neurocomputing **620**, 129267 (2025)
3. Nachaoui, M., Laghrib, A., Hakim, M.: A new space-variant optimization approach for image segmentation. In: International Conference on Numerical Analysis and Optimization Days, pp. 87–97 (2021). Springer
4. Ourdou, A., Ghazdali, A., Laghrib, A., Metrane, A.: Blind separation of instantaneous mixtures of independent/dependent sources. Circuits Syst. Signal Process. **40**(9), 4428–4451 (2021)
5. Ghazdali, A., Hakim, A., Laghrib, A., Mamouni, N., Raghay, S.: A new method for the extraction of fetal ECG from the dependent abdominal signals using blind source separation and adaptive noise cancellation techniques. Theor. Biol. Med. Model. **12**, 1–20 (2015)

6. Alexander, R.M.: Optimum take-off techniques for high and long jumps. Philos. Trans. R. Soc. London Ser. B Biol. Sci. **329**(1252), 3–10 (1990)
7. Panoutsakopoulos, V., Papaiakovou, G., Katsikas, F., Kollias, I.: 3D biomechanical analysis of the preparation of the long jump take-off. New Stud. Athletics **25**, 55–68 (2010)
8. Lazzari, B.D., Mascia, G., Vannozzi, G., Camomilla, V.: Estimating the standing long jump length from smartphone inertial sensors through machine learning algorithms. Bioengineering **10**(5), 546 (2023). https://doi.org/10.3390/bioengineering10050546
9. Olympics Official Website. https://olympics.com/en/. Accessed 24 Oct 2024
10. Willmott, C.J., Matsuura, K.: Advantages of the mean absolute error (MAE) over the root mean square error (RMSE) in assessing average model performance. Climate Res. **30**(1), 79–82 (2005)
11. Chai, T., Draxler, R.R.: Root mean square error (RMSE) or mean absolute error (MAE)? - arguments against avoiding RMSE in the literature. Geosci. Model Dev. **7**(3), 1247–1250 (2014)
12. The use and interpretation of regression coefficients and R^2. Int. J. Qual. Reliab. Manag. **23**(6), 620–628 (2006)
13. Breiman, L.: Random forests. Mach. Learn. **45**(1), 5–32 (2001). https://doi.org/10.1023/A:1010933404324
14. Chen, T., Guestrin, C.: Xgboost: a scalable tree boosting system. In: Proceedings of the 22nd ACM SIGKDD International Conference on Knowledge Discovery and Data Mining, pp. 785–794. ACM (2016). https://doi.org/10.1145/2939672.2939785
15. Prokhorenkova, L., Gusev, G., Vorobev, A., Dorogush, A.V., Gulin, A.: Catboost: unbiased boosting with categorical features. In: Advances in Neural Information Processing Systems 31 (NeurIPS 2018) (2018)
16. Friedman, J.H.: Greedy function approximation: a gradient boosting machine. Ann. Stat. **29**(5), 1189–1232 (2000). https://doi.org/10.1214/aos/1013203451
17. Khoudi, Z., Nachaoui, M., Lyaqini, S.: Identifying the contextual factors related to the reading performance of moroccan fourth-grade students from a machine learning-based approach. Educ. Inf. Technol. **29**(3), 3047–3073 (2024)
18. Khoudi, Z., Nachaoui, M., Lyaqini, S.: Finding the contextual impacts on students' mathematical performance using a machine learning-based approach. Infocommun. J. **16**(SI), 12–21 (2024)

A Stochastic Primal-Dual Fixed Point Approach for Image Super-Resolution

Yasmine El Mobariki[1(✉)], Amine Laghrib[2], and Abdelghani Ghazdali[1]

[1] Lab LIPIM, Sultan Moulay Slimane University, ENSA, Khouribga, Morocco
Yasmine.elmobariki@usms.ac.ma
[2] EMI, Sultan Moulay Slimane University, Beni Mellal, Morocco

Abstract. In this paper, we propose a Stochastic Primal-Dual Fixed Point (SPDFP) approach, leveraging to address the challenges of super-resolution reconstruction. Unlike traditional full-gradient methods, SPDFP employs stochastic gradients making it particularly well-suited for handling high-dimensional image data. Through numerical experiments, we demonstrate the superiority of our approach in enhancing resolution while preserving fine details, outperforming existing methods in both efficiency and reconstruction quality. Beyond empirical validation, we establish a solid theoretical framework, analyzing convergence properties under specific assumptions. This deeper insight into the method's behavior highlights its practical effectiveness and sheds light on its potential and limitations in real-world super-resolution tasks.

Keywords: Total variation · convex optimization · proximity operator · Super-Resolution

1 Introduction

When an image is captured under suboptimal conditions or through an imperfect optical system, the resulting observation often suffers from low resolution, loss of fine details, and degraded visual quality. The goal of multiframe image super-resolution (MISR) is to reconstruct a high-resolution image by fusing multiple low-resolution frames, leveraging their complementary information to enhance spatial details and restore image fidelity [1].

Super-resolution techniques play a crucial role in various domains, including pattern recognition and machine intelligence, where high-quality images are essential for accurate analysis and decision-making. Mathematically, multiframe super-resolution shares a common framework with other inverse imaging problems. For instance, in medical imaging, modalities such as computed tomography (CT), proton computed tomography (pCT), and magnetic resonance imaging (MRI) often operate under constraints imposed by limited data acquisition. These challenges arise due to equipment scarcity and patient availability, leading to incomplete or noisy measurements. As a result, advanced computational

A. Laghrib and A. Ghazdali—These authors contributed equally to this work.

A. Laghrib and A. Ghazdali (Eds.): ICMDS 2024, LNNS 1466, pp. 36–51, 2025.
https://doi.org/10.1007/978-3-031-94802-2_4

techniques are required to extract meaningful high-resolution information from imperfectly acquired images [2].

In the context of multiframe super-resolution, we assume the availability of multiple observed frames $\{Y_i\}_{i=1}^N$, each of which is degraded by a slightly different blurring, decimation and transformations operators H_i, capturing variations due to motion, perspective changes, or sensor noise. The generalized forward model is then given by:

$$H_i X + \eta_i = Y_i, \quad i = 1, \dots, N. \tag{1}$$

The objective of multiframe super-resolution is to reconstruct X by leveraging the information across multiple frames, thereby enhancing spatial resolution beyond the limitations of individual observations.

1.1 Related Works

There are numerous methods in the literature dedicated to the problem of super-resolution. Broadly speaking, these techniques can be categorized into two main groups:

- Model-Based Approaches: These methods exploit a priori knowledge of the degradation process and often incorporate regularization strategies to address the inherent ill-posedness of the inverse problem. Classical techniques such as iterative back-projection, Tikhonov regularization, and variational methods fall under this category [3–5]. They typically assume a known blur kernel and noise statistics, thereby facilitating the derivation of explicit reconstruction formulas or iterative schemes [6–8].
- Learning-Based Approaches: With the advent of deep learning, methods based on convolutional neural networks (CNNs) and generative adversarial networks (GANs) have become popular. These techniques learn an end-to-end mapping from low-resolution to high-resolution images using large-scale datasets [9,10]. While they have demonstrated impressive performance on challenging real-world images, their reliance on extensive training data and high computational costs can be prohibitive in certain applications [11–13].

Despite the significant advances achieved by these approaches, a fundamental trade-off remains between interpretability and computational efficiency. Model-based methods offer clear theoretical insights and controllability but may struggle with the high complexity of real-world data. In contrast, while learning-based methods often deliver superior visual quality, they can lack transparency and require substantial computational resources.

There is many method in literature that are dedicated to the problem of super resolution, To recover X, we formulate a regularized optimization problem that balances data fidelity with prior information about X:

$$\min_{X \in L^2(\Omega)} \left\{ \sum_{i=1}^{N} \frac{1}{2} \|H_i X - Y_i\|_2^2 + \lambda R(X) \right\}, \tag{2}$$

where:

- The data fidelity term $\frac{1}{2}\|H_iX - Y_i\|_2^2$ enforces consistency with the observed frames.
- $R(X)$ is a regularization functional promoting desirable features in X; for example, one may choose Total Variation (TV) regularization [14–16]:

$$R(X) = \|\nabla X\|_1,$$

  which is effective at preserving edges.
- $\lambda > 0$ is a parameter balancing the fidelity and regularization terms.

The classical variational approach has been extensively used in image processing and inverse problems. However, when applied to multiframe super-resolution, several challenges arise:

- Computational Complexity: Super-resolution tasks typically deal with high-dimensional data and multiple frames, leading to large-scale optimization problems. The iterative methods and numerical approximations required can be computationally demanding, especially when aiming for high-quality reconstructions.
- Sensitivity to Initial Conditions: Variational models for super-resolution are highly sensitive to initial conditions. The choice of initial parameters or starting points can significantly impact the final high-resolution reconstruction, necessitating careful initialization and parameter tuning.
- Ambiguity in Parameter Selection: These models often involve various parameters such as regularization weights, penalty terms, and other optimization settings. Striking an optimal balance between data fidelity and regularization remains challenging and can be subjective.
- Dependence on Expert Knowledge: Classical variational approaches generally require substantial manual intervention and expert knowledge to guide the optimization process and select appropriate parameters, limiting the degree of automation.
- Limited Capability for Nonlinear and Complex Models: Traditional variational methods are typically designed for linear or simple models. In the context of super-resolution, where the relationship between the observed low-resolution frames and the high-resolution image can be highly nonlinear and complex, these approaches may struggle to capture intricate details.

To address these challenges, researchers have been integrating advanced techniques into the variational framework for super-resolution. Such enhancements include:

- Adaptive Regularization: By dynamically adjusting the regularization parameters, adaptive methods help control the complexity of the trial function, providing a better trade-off between reconstruction accuracy and computational cost[1, 17].
- Advanced Optimization Algorithms: The use of sophisticated optimization methods, such as stochastic optimization, accelerated proximal methods, and primal-dual algorithms with momentum (e.g., Nesterov's acceleration),

can significantly improve convergence properties and computational efficiency [18].
- Machine Learning-Based Techniques: Incorporating data-driven approaches, such as deep neural networks, into the variational framework enables the design of flexible trial functions that can learn complex, nonlinear relationships inherent in multiframe super-resolution problems [19].

By combining these advanced techniques with the classical variational approach, the overall performance and robustness in solving challenging multiframe super-resolution tasks can be substantially improved [20].

1.2 Proposed Approach

In this work, we advocate a hybrid strategy that integrates the strengths of both paradigms. Specifically, we introduce a Stochastic Primal-Dual Fixed Point (SPDFP) algorithm, inspired from the work [21], tailored for multiframe super-resolution. Our method leverages stochastic gradient updates to efficiently process multiple low-resolution frames and incorporates a robust regularization framework to mitigate the effects of noise and blur. Let $\{X^{(k)}, P^{(k)}\}$ denote the estimates at iteration k. The accelerated SPDFP algorithm performs the following updates:

$$\begin{aligned} P^{(k+1)} &= \operatorname{prox}_{\sigma\delta_{\mathcal{P}}^*}\Big(P^{(k)} + \sigma K\bar{X}^{(k)}\Big), \\ X^{(k+1)} &= \operatorname{prox}_{\tau G}\Big(X^{(k)} - \tau\Big(K^* P^{(k+1)} + \sum_{i=1}^{N} H_i^*(H_i X^{(k)} - Y_i)\Big)\Big), \\ \bar{X}^{(k+1)} &= X^{(k+1)} + \theta\Big(X^{(k+1)} - X^{(k)}\Big), \end{aligned} \tag{3}$$

where:

- τ and σ are step-size parameters,
- $\theta \in [0, 1]$ is the momentum parameter,
- K^* and H_i^* are the adjoint operators of K (linear regularization operator) and H_i, respectively,
- $\operatorname{prox}_{\tau G}$ denotes the proximal operator corresponding to the function G (which encapsulates the data fidelity and regularization terms),
- $\operatorname{prox}_{\sigma\delta_{\mathcal{P}}^*}$ is the proximal operator for the convex conjugate of the indicator function.

To solve (2) efficiently, especially for large-scale multiframe data, we adopt a SPDFP method. A saddle-point formulation of the problem can be written as:

$$\min_{X\in L^2(\Omega)} \max_{P\in\mathcal{P}} \left\{ \sum_{i=1}^{N} \frac{1}{2}\|H_i X - Y_i\|_2^2 + \langle KX, P\rangle - \delta_{\mathcal{P}}(P) + \lambda R(X) \right\},$$

where:

- K is a linear operator associated with the regularization (e.g., the gradient operator when using TV regularization).
- P denotes the dual variable.
- $\delta_{\mathcal{P}}(P)$ is the indicator function of the dual feasible set $\mathcal{P}$.

A typical stopping criterion is based on the relative change in X:

$$\text{RelErr}^{(k)} = \frac{\|X^{(k+1)} - X^{(k)}\|_2}{\|X^{(k)}\|_2} < \epsilon,$$

where $\epsilon > 0$ is a small tolerance level.
The proposed SPDFP algorithm, is well-suited for multiframe super-resolution tasks. By leveraging stochastic updates, the algorithm efficiently handles multiple low-resolution observations, thereby accelerating convergence. The incorporation of a regularization term such as TV further ensures that the reconstructed image X retains important structural features, while the overall framework guarantees that both data fidelity and prior knowledge are effectively balanced.

The remainder of this paper is structured as follows:

- Section 2 presents the mathematical formulation of the multiframe super-resolution problem, including the regularized optimization framework and the derivation of the accelerated SPDFP algorithm.
- Section 3 provides extensive numerical experiments that compare the performance of our approach against state-of-the-art methods.
- Section 4 concludes the paper and outlines directions for future research.

2 Proposed Stochastic Primal Dual Fixed-Point Method

In multiframe super-resolution, the goal is to reconstruct a high-resolution image from multiple low-resolution observations affected by blurring and noise. The reconstruction process seeks to minimize a cost function that consists of two components: a data fidelity term, ensuring consistency with observed measurements, and a regularization term, which imposes constraints to promote desirable properties such as smoothness or sparsity.

Instead of computing the full gradient of the objective function, a stochastic gradient approach is employed, where the gradient is estimated using a random subset of the data. This stochastic estimate serves as an unbiased approximation of the true gradient, enabling efficient optimization while handling large datasets. Consider a bounded domain Ω with sufficient regularity on its boundary. Let $f_i \in L^2(\Omega)$ $(i = 1, \ldots, n)$ denote a set of observed low-resolution images that correspond to different frames of the same scene, where each frame is a degraded version of an unknown high-resolution image X. The degradation process is modeled as follows:

$$H_i X + \eta_i = f_i, \quad i = 1, \ldots, n, \tag{4}$$

where:

- $H_i \in L^\infty(\Omega)$ represents the degradation operator for the i-th frame, which typically accounts for blurring and downsampling,
- η_i represents the additive noise affecting each observation.

To recover the high-resolution image X, we formulate a regularized inverse problem:

$$X^* = \underset{X}{\operatorname{argmin}} \sum_{i=1}^{n} \|H_i X - f_i\|_2^2 + \nu \|DX\|_1, \tag{5}$$

where:

- The term $\sum_{i=1}^{n} \|H_i X - f_i\|_2^2$ enforces consistency with observed data across multiple frames,
- The regularization term $\nu\|DX\|_1$ imposes sparsity on the image gradients to preserve edges while reducing noise, where $\nu > 0$ is a regularization parameter.

In practice, the high-resolution image X is represented on an $m \times n$ Cartesian grid and vectorized using lexicographic ordering. The discrete gradient operator is defined as:

$$DX = \begin{bmatrix} D_x \\ D_y \end{bmatrix} X, \tag{6}$$

where D_x and D_y are finite difference operators with periodic boundary conditions, approximating horizontal and vertical derivatives, respectively.

The space of image vectors is denoted as $\mathcal{X} = \mathbb{R}^M$, where $M = mn$ is the total number of pixels. The gradient belongs to the space $\mathcal{Y} = \mathbb{R}^{2M}$, where standard norms apply. In particular, the L_1 norm of the gradient:

$$\|DX\|_1 = \sum_{p=1}^{M} \left(|(D_x X)_p| + |(D_y X)_p|\right), \tag{7}$$

is commonly referred to as TV regularization. An alternative formulation known as isotropic total variation is given by the $L_{2,1}$ norm:

$$\|DX\|_{2,1} = \sum_{p=1}^{M} \sqrt{(D_x X)_p^2 + (D_y X)_p^2}. \tag{8}$$

In this work, we employ anisotropic TV regularization, though our approach can be extended to other regularization norms. For convenience, let us introduce the following notation:

$$g_1(X) = \nu\|X\|_1,$$
$$g_2(X) = \sum_{i=1}^{n} \|H_i X - f_i\|_2^2.$$

With these definitions, the optimization problem can be rewritten as:

$$X^* = \underset{X \in \Omega}{\operatorname{argmin}} \quad g_2(X) + (g_1 \circ D)(X). \tag{9}$$

The formulation in (9) follows a convex variational framework, where the first term ensures consistency with multiframe observations, while the second term imposes structured sparsity via total variation. This formulation naturally lends itself to efficient numerical optimization techniques, such as primal-dual methods, proximal algorithms, and accelerated stochastic gradient approaches.

The Primal-Dual Fixed Point method is a widely used optimization algorithm for solving minimization problems [22]. It is particularly effective for both convex and non-convex optimization challenges, leveraging an iterative approach to update both primal and dual variables until convergence is reached. At each iteration, the primal variables—representing the solution—are updated alongside the dual variables, which account for the problem's constraints. This duality framework enables PDFP to efficiently exploit the problem structure, often resulting in accelerated convergence compared to alternative techniques. Moreover, its robustness in handling large-scale optimization problems stems from its ability to incorporate various structural constraints. Over time, numerous adaptations and extensions have been developed to enhance its efficiency and broaden its applicability across different optimization domains. As a result, PDFP has become a powerful and versatile tool in scientific, engineering, and computational disciplines for addressing complex minimization problems.

2.1 Primal Dual Fixed Point Method

The key idea behind the Primal-Dual Fixed Point method is to iteratively update primal and dual variables until convergence to the optimal solution. The associated PDFP algorithm for the minimization problem (9) is then given by:

$$\begin{cases} V^* & = (I - \text{Prox}_{\frac{\gamma}{\lambda} g_1})(D(X^* - \gamma \nabla g_2(X)) + (I - \lambda D D^T)V^*) \\ & = \widetilde{\psi_0}(X^*, V^*). \\ X^* & = X^* - \gamma \nabla g_2(X^*) - \lambda D^T . \widetilde{\psi_0}(X^*, V^*). \end{cases} \tag{10}$$

Then the PDFP is given as:

Algorithm : Primal Dual Fixed Point Method

- Set $X_1 \in \mathbb{R}^2$, $V_1 \in \mathbb{R}^2$ and choose $\gamma > 0$ and $\lambda > 0$ then,
- For $k = 1, 2, \ldots$

$$\begin{aligned} X_{k+\frac{1}{2}} &= X_k - \gamma \nabla g_2(X_k) \\ V_{k+1} &= (I - Prox_{\frac{\gamma}{\lambda} g_1})(D X_{k+\frac{1}{2}} + (I - \lambda D D^T)V_k) \\ X_{k+1} &= X_{k+\frac{1}{2}} - \lambda D^T V_{k+1} \end{aligned}$$

Until the stopping criterion is satisfied

The abundance of information in video image processing naturally lends itself to the use of stochastic techniques, which are well-suited for handling the inherent complexities and uncertainties in dynamic visual data. Unlike traditional deterministic approaches, which may struggle to accurately capture the variability and noise present in video sequences, stochastic methods provide a flexible framework for robust processing. By leveraging stochastic strategies, we can effectively utilize the available information to enhance the quality of reconstructed images and improve the overall performance of image processing in dynamic environments. Given that $g_2(X) = \sum_{i=1}^{n} \|H_i X - f_i\|_2^2$, its gradient is expressed as

$$\nabla g_2(X) = \sum_{i=1}^{n} 2H_i^T(H_i X - f_i). \tag{11}$$

Building on this formulation, we now introduce the stochastic version of the Primal-Dual Fixed Point method.

2.2 Stochastic Primal Dual Fixed Point Method

In the stochastic variant of the Projected Direct Full-Newton Step approach, the full gradient used in each step is replaced with a stochastic gradient. This adaptation is particularly useful for handling enormous large-scale problems where computing the full gradient may be impractical or computationally expensive. This configuration is realized in two steps:

- Given batch size p, the set $\{1, 2, \ldots, n\}$ is divided into $\frac{n}{p}$ non-overlapping batches, where the number of entries of each batch is p.
- Then, the batch is chosen with probability p and the stochastic gradient of g_2 at X denoted by $\nabla g_i^{[i]}(X)$ is given by:

$$\nabla g_2^{[i]}(X) = \frac{1}{p} \sum_{j=(i-1)p+1}^{ip} 2H_j^T(H_j X - f_j)$$

We can verify that,

$$\mathbb{E}(\nabla g_2^{[i]}) = \nabla g_2(X)$$

In a stochastic setting, the fixed-point formulation must be adjusted. We need a Lemma to extend the PDFP Algorithm. The following Lemma can be directly derived from ([21], Lemma 3.2):

Lemma 1. *Let $\lambda > 0, \gamma_k > 0$ be two positive numbers.*
Supposing that X^* is a solution of (5), then there exists $V^* \in \partial g_1(DX^*)$ such that

$$\begin{aligned} V^* &= (I - Prox_{\phi_k})(\frac{\lambda}{\gamma_k} D(X^* - \gamma_k \nabla g_2(X^*)) + (I - \lambda DD^T)V^*) \\ &= \Gamma_0^{(k)}(X^*, V^*) \\ X^* &= X^* - \gamma_k \nabla g_2(X^*) - \gamma_k D^T \circ \Gamma_0^{(k)}(X^*, V^*) \end{aligned}$$

where

$$\phi_k(x) = \frac{\lambda}{\gamma_k} g_1(\frac{\gamma_k}{\lambda} x).$$

Conversely, if X^*, V^* satisfy the above equations then X^* is the solution of (5).

The algorithm we propose is formalized as Algorithm 1.

Algorithm 1 : Stochastic Primal Dual Fixed Point Method

- Set $X_1 \in \mathbb{R}^2$, $V_1 \in \mathbb{R}^2$ and choose c, λ, $\alpha \in]0.1]$
- For $k = 1, 2, \ldots$
  $\gamma_k = \dfrac{c}{k^\alpha}$
  Choose i_k randomly from $1, 2, \ldots, \frac{n}{p}$

$$X_{k+\frac{1}{2}} = X_k - \gamma_k \nabla g_2^{[i_k]}(X_k)$$
$$V_{k+1} = \frac{\lambda}{\gamma_k}(I - Prox_{\frac{\gamma_k}{\lambda} g_1})(DX_{k+\frac{1}{2}} + (I - \lambda DD^T)\frac{\gamma_k}{\lambda} V_k)$$
$$X_{k+1} = X_{k+\frac{1}{2}} - \gamma_k D^T V_{k+1}$$

Until the criterion is satisfied.

Let $V_{k+1} = \frac{\gamma_k}{\lambda} V_{k+1}$ we obtain the equivalent form :

In the following section, we present some numerical results to confirm the choice of the SPDFP algorithm.

3 Numerical Part

In this section, we conduct a series of extensive experiments to assess and validate the effectiveness of the proposed optimization approach for multiframe super-resolution reconstruction. We begin by outlining the numerical settings and proceed with simulated tests to demonstrate the accuracy and robustness of our inverse problem formulation. Subsequently, we evaluate the performance of our method on real-world low-resolution image sequences, considering both Gaussian noise and unknown real-world degradations. A comparative analysis is carried out against several recent state-of-the-art super-resolution techniques, including the Combined High-Order Regularization (CHOR) [7], Modified Second Order Method (MSOM) [8], and the Spatially Varying Regularization (SVR) [6]. The parameters for these methods are selected according to the configurations provided in their respective studies.

For all the following experiments, we employed MATLAB R2015 on a laptop equipped with an Intel(R) Core i5 CPU with 2.30 GHz. To quantitatively evaluate the performance of the reconstructed high-resolution images, we use the Peak

Signal-to-Noise Ratio (PSNR) as a primary metric, which measures the fidelity of the super-resolved images. Additionally, we incorporate the Structural Similarity Index (SSIM) to complement the PSNR values, offering a perceptual assessment of image quality and structural consistency.

We conduct two simulated tests to evaluate the effectiveness of the proposed multiframe super-resolution approach under different levels of Gaussian blur. A decimation factor of 4 is applied to all tests, simulating the resolution degradation typically encountered in practical scenarios. For the first test:

- **Test Image:** We use the *City* image as the reference high-resolution image for evaluation.
- **Blur Kernel:** A Gaussian blurring kernel with a size of 5×5 and a standard deviation of 1.5 is applied to simulate degradation.
- **Decimation Factor:** A downsampling operation with a factor of 4 is performed on all images, reducing the resolution before reconstruction.
- **Noise Level:** A Gaussian noise is added with variance $\sigma^2 = 0.01$.
- **Parameters:** A step size of $1e - 5$ is chosen to ensure stable and efficient convergence in the iterative optimization process. $\lambda = 0.1$, $\alpha = 0.2$, $\nu = 0.01$, maximum iteration $N = 1000$ and $c = 1$ for our Algorithm.

To compare the reconstruction quality, we evaluate the performance of the three state-of-the-art super-resolution methods along with the proposed approach. The visual results of the different methods are presented in Fig. 1, providing a direct comparison of the reconstructed high-resolution images. From these observations, we conclude that the proposed method outperforms existing approaches, achieving higher image quality by effectively suppressing blur and reconstructing fine details. Concerning the second test, we aim to see the robustness of our Algorithm against blur reduction:

- **Test Image:** We use the *Butterfly* image as the reference high-resolution image for evaluation.
- **Blur Kernel:** A Gaussian blurring kernel with a size of 9×9 and a standard deviation of 1.5 is applied to simulate degradation.
- **Decimation Factor:** A downsampling operation with a factor of 4 is performed on all images, reducing the resolution before reconstruction.
- **Noise Level:** A Gaussian noise is added with variance $\sigma^2 = 0.01$.
- **Parameters:** A step size of $1e - 4$. $\lambda = 0.08$, $\alpha = 0.1$, $\nu = 0.01$, maximum iteration $N = 1000$ and $c = 1$ for our Algorithm.

The visual outcomes of the various methods are displayed in Fig. 2, offering a clear comparison of the reconstructed high-resolution images. Based on these observations, we conclude that the proposed method surpasses existing techniques, delivering superior image quality by efficiently mitigating blur and preserving intricate details. To further validate the visual comparison, we present a quantitative analysis using the PSNR and SSIM metrics in Tables 1 and 2, respectively. We randomly selected six images from the *Train* dataset and applied various blur and noise conditions to them. These conditions include different blur

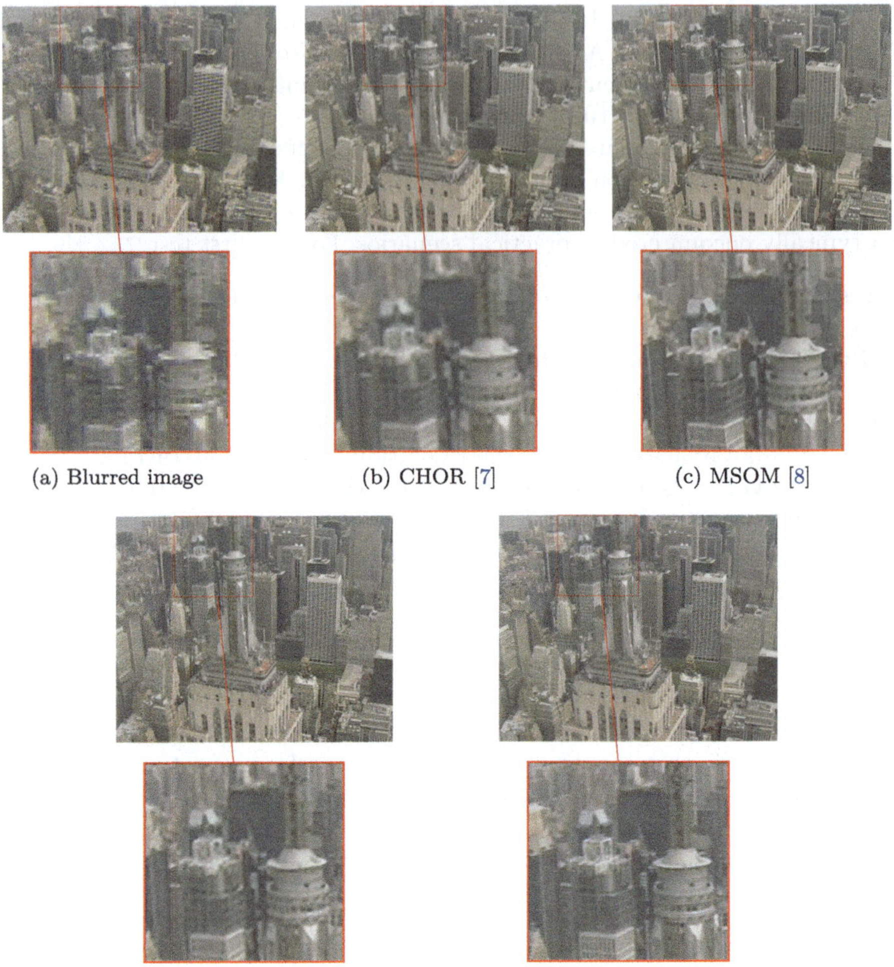

(a) Blurred image (b) CHOR [7] (c) MSOM [8]

(d) SVR [6] (e) Proposed model

Fig. 1. Comparison between some super-resolution methods and the proposed model for the (*City* image).

sizes and Gaussian noise levels, chosen to simulate a range of real-world image degradation scenarios. The resulting images were then used to evaluate and compare the performance of the proposed super-resolution method against existing approaches. These measurements provide a more objective evaluation of image quality, reinforcing the superior performance of the proposed method.

In a last real test scenario, we turn our attention to the *Blue* image. This particular image has been deliberately subjected to a more intense form of corruption and we aim to augment it resolution by a factor of 4. The recovered clean images using the proposed method and compared to other super-resolution algo-

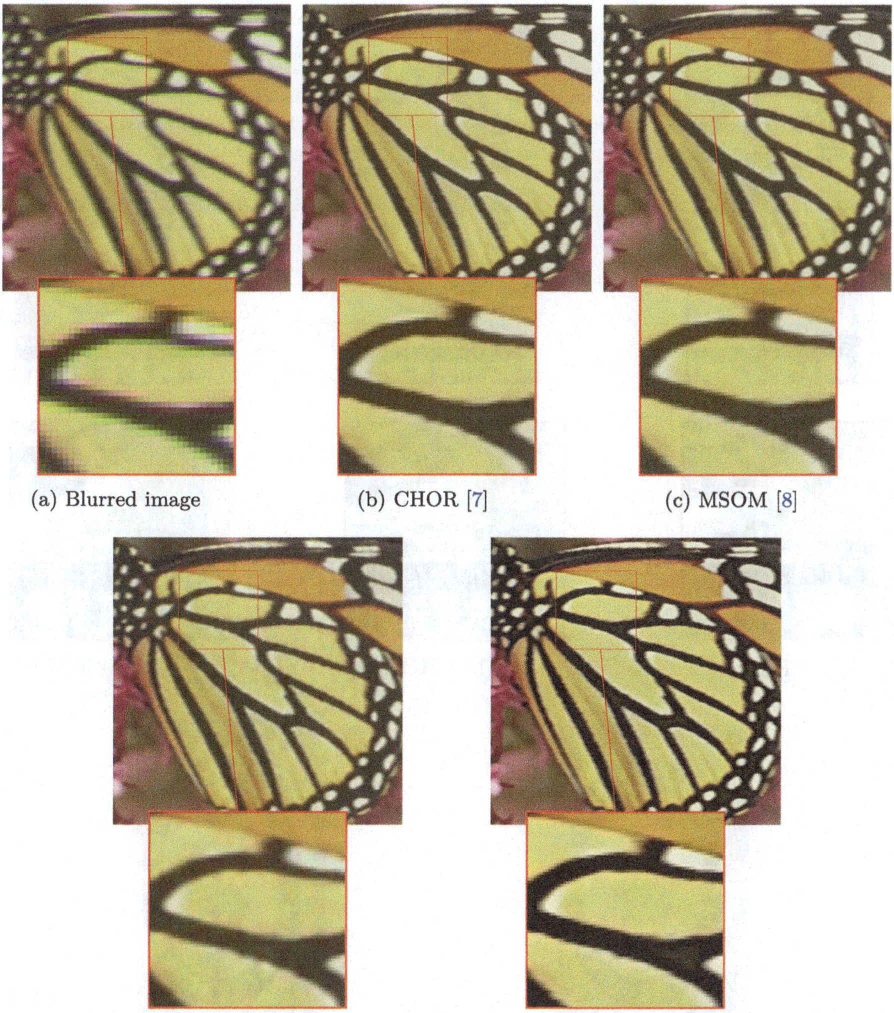

(a) Blurred image (b) CHOR [7] (c) MSOM [8]

(d) SVR [6] (e) Proposed model

Fig. 2. Comparison between some super-resolution methods and the proposed model for the (*Butterfly* image).

rithms are depicted in Fig. 3. Once again, and in the presence of high blur, the proposed approach achieves better reconstruction result.

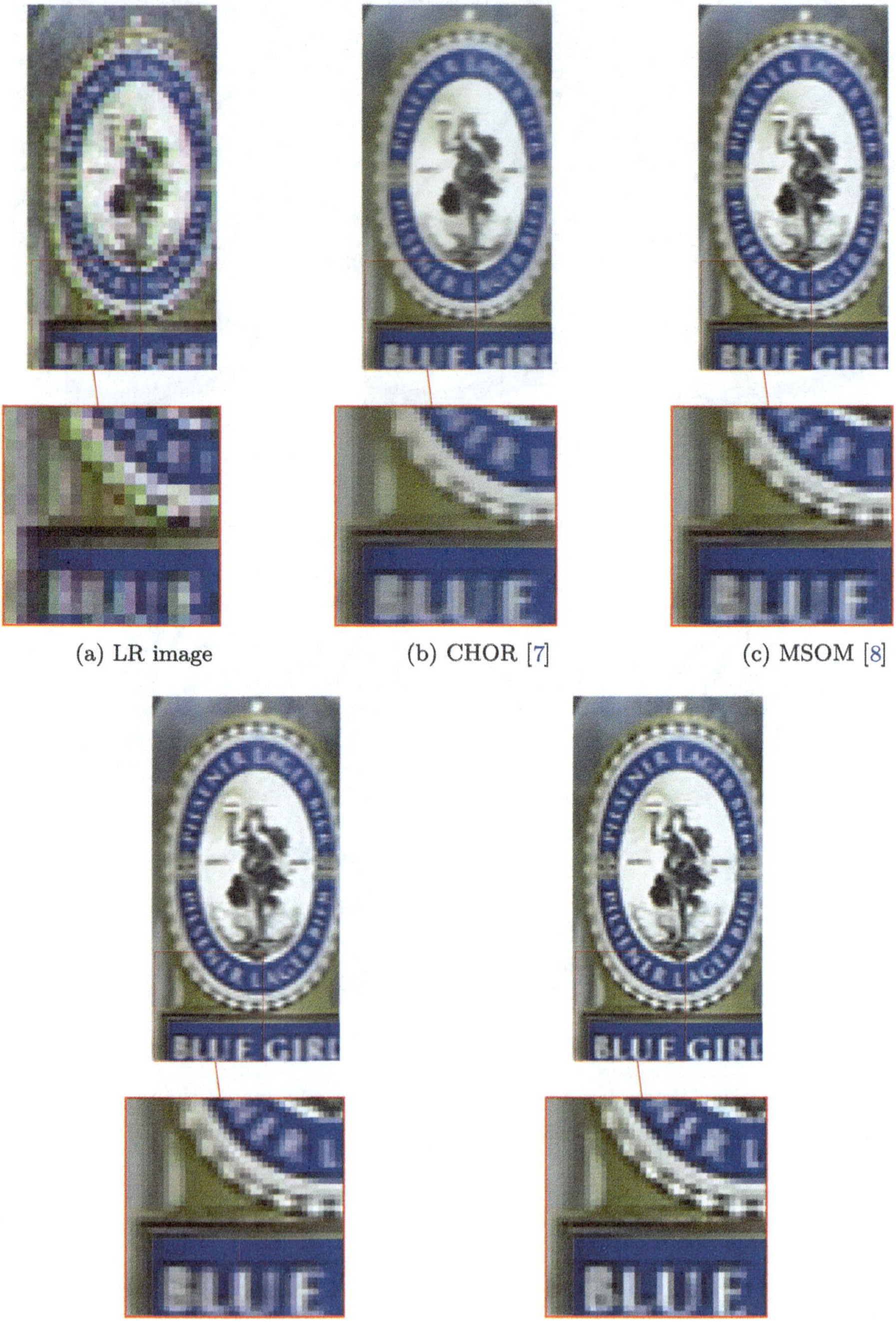

(a) LR image (b) CHOR [7] (c) MSOM [8]

(d) SVR [6] (e) Proposed model

Fig. 3. Comparison between some super-resolution methods and the proposed model for the real (*Blue* image).

Table 1. Quantitative Comparison of PSNR for Different Methods on Selected Test Images for Super-Resolution (Including Blur Size and Gaussian Noise).

Test Image	Blur Size	Gaussian Noise	CHOR [7]	MSOM [8]	SVR [6]	Proposed Method
Star	5×5	$\sigma = 15$	PSNR: 28.50	PSNR: 29.10	PSNR: 30.20	**PSNR: 32.15**
Woman	7×7	$\sigma = 20$	PSNR: 25.75	PSNR: 26.30	PSNR: 27.45	**PSNR: 29.00**
Tiger 1	3×3	$\sigma = 10$	PSNR: 31.40	PSNR: 32.00	PSNR: 33.10	**PSNR: 34.25**
Tiger 2	5×5	$\sigma = 12$	PSNR: 27.10	PSNR: 27.60	PSNR: 28.75	**PSNR: 30.50**
Festival	9×9	$\sigma = 18$	PSNR: 29.20	PSNR: 29.80	PSNR: 31.00	**PSNR: 32.60**
Bear	7×7	$\sigma = 14$	PSNR: 26.40	PSNR: 27.20	PSNR: 28.30	**PSNR: 29.80**

Table 2. Quantitative Comparison of SSIM for Different Methods on Selected Test Images for Super-Resolution (Including Blur Size and Gaussian Noise).

Test Image	Blur Size	Gaussian Noise	CHOR [7]	MSOM [8]	SVR [6]	Proposed Method
Star	5×5	$\sigma = 15$	SSIM: 0.850	SSIM: 0.860	SSIM: 0.880	**SSIM: 0.910**
Woman	7×7	$\sigma = 20$	SSIM: 0.820	SSIM: 0.830	SSIM: 0.840	**SSIM: 0.870**
Tiger 1	3×3	$\sigma = 10$	SSIM: 0.880	SSIM: 0.890	SSIM: 0.900	**SSIM: 0.920**
Tiger 2	5×5	$\sigma = 12$	SSIM: 0.800	SSIM: 0.810	SSIM: 0.830	**SSIM: 0.850**
Festival	9×9	$\sigma = 18$	SSIM: 0.840	SSIM: 0.850	SSIM: 0.860	**SSIM: 0.890**
Bear	7×7	$\sigma = 14$	SSIM: 0.780	SSIM: 0.790	SSIM: 0.810	**SSIM: 0.830**

4 Conclusions

In this paper, we propose a Stochastic Primal-Dual Fixed Point (SPDFP) approach to tackle the challenges inherent in super-resolution reconstruction. Unlike conventional full-gradient methods, SPDFP leverages stochastic gradients, making it particularly advantageous for high-dimensional image data. Through extensive numerical experiments, we demonstrate the superior performance of our approach, achieving significant improvements in both resolution enhancement and fine detail preservation, surpassing existing methods in terms of efficiency and reconstruction quality. Beyond the empirical evidence, we also provide a robust theoretical foundation, analyzing the convergence properties of the method under certain assumptions. This theoretical framework not only highlights the practical effectiveness of SPDFP but also offers a deeper understanding of its behavior, revealing both its strengths and potential limitations in real-world super-resolution tasks.

Looking ahead, the integration of machine learning techniques offers exciting opportunities to further optimize the configuration of model parameters. By leveraging data-driven approaches, such as deep learning-based parameter tuning or reinforcement learning, we can potentially automate the selection of optimal settings, improving adaptability across diverse datasets. Additionally, the SPDFP approach holds promise for broader applications, including image denoising, compression, and even medical image analysis, where precise resolution enhancement is critical. This opens up a wealth of future possibilities for advancing image processing techniques and their integration into practical, real-world scenarios.

Data Availability Statement. The datasets employed in this research can be obtained by contacting the corresponding author and submitting a reasonable request.

Conflict of Interest. All the above authors approve that they don't have any conflict of interest.

References

1. Laghrib, A., Bella, F., Nachaoui, M., Hakim, M.: Learning nonlocal weights for second-order nonlocal super-resolution. Discrete Continuous Dyn. Syst.-S **18**(1), 238–268 (2025)
2. Idriss, E.M., Hakim, A., Laghrib, A.: On the well-posedness of a tensor-based second order PDE with bilateral term for image super-resolution. Evol. Equ. Control Theory **12**(2) (2023)
3. Nachaoui, M., Laghrib, A.: An improved bilevel optimization approach for image super-resolution based on a fractional diffusion tensor. J. Franklin Inst. **359**(13), 7165–7195 (2022)
4. Laghrib, A., Hadri, A., Hakim, A.: An edge preserving high-order PDE for multiframe image super-resolution. J. Franklin Inst. **356**(11), 5834–5857 (2019)
5. Laghrib, A., Chakib, A., Hadri, A., Hakim, A.: A nonlinear fourth-order PDE for multi-frame image super-resolution enhancement. Discrete Continuous Dyn. Syst.-B **25**(1), 415 (2020)
6. Hadri, A., Khalfi, H., Laghrib, A., Nachaoui, M.: An improved spatially controlled reaction-diffusion equation with a non-linear second order operator for image super-resolution. Nonlinear Anal. Real World Appl. **62**, 103352 (2021)
7. Laghrib, A., Ezzaki, M., El Rhabi, M., Hakim, A., Monasse, P., Raghay, S.: Simultaneous deconvolution and denoising using a second order variational approach applied to image super resolution. Comput. Vis. Image Underst. **168**, 50–63 (2018)
8. Laghrib, A., Aitbella, F., Hakim, A.: A second-order nonlocal regularized variational model for multiframe image super-resolution. Int. J. Nonlinear Sci. Numer. Simul. **23**(3–4), 385–399 (2022)
9. Huang, S., Wang, J., Yang, Y., Wan, W.: Tarn: a lightweight two-branch adaptive residual network for image super-resolution. Int. J. Mach. Learn. Cybern. **15**(9), 4119–4132 (2024)
10. Mishra, S.R., Mohapatra, H., Saxena, S.: Leveraging data analytics and a deep learning framework for advancements in image super-resolution techniques: From classic interpolation to cutting-edge approaches. In: Data Analytics and Machine Learning: Navigating the Big Data Landscape, pp. 105–126. Springer (2024)
11. Pang, K., Zhao, K., Hung, A., Zheng, H., Yan, R., Sung, K.: Nexpr: neural explicit representation for fast arbitrary-scale medical image super-resolution. Comput. Biol. Med. **184**, 109354 (2025)
12. Zhao, L., Chi, H., Zhong, T., Jia, Y.: Perception-oriented generative adversarial network for retinal fundus image super-resolution. Comput. Biol. Med. **168**, 107708 (2024)

13. Zhong, T., Yang, F., Dong, X., Dong, S., Luo, Y.: Shbgan: hybrid bilateral attention GAN for seismic image super-resolution reconstruction. IEEE Trans. Geosci. Remote Sens. (2024)
14. Mitzel, D., Pock, T., Schoenemann, T., Cremers, D.: Video super resolution using duality based TV-L 1 optical flow. In: Pattern Recognition: 31st DAGM Symposium, Jena, Germany, 9–11 September 2009. Proceedings 31, pp. 432–441. Springer (2009)
15. Marquina, A., Osher, S.J.: Image super-resolution by TV-regularization and bregman iteration. J. Sci. Comput. **37**, 367–382 (2008)
16. Laghrib, A., Ghazdali, A., Hakim, A., Raghay, S.: A multi-frame super-resolution using diffusion registration and a nonlocal variational image restoration. Comput. Math. Appl. **72**(9), 2535–2548 (2016)
17. El Mourabit, I., El Rhabi, M., Hakim, A., Laghrib, A., Moreau, E.: A new denoising model for multi-frame super-resolution image reconstruction. Signal Process. **132**, 51–65 (2017)
18. Nachaoui, M., Afraites, L., Laghrib, A.: A regularization by denoising super-resolution method based on genetic algorithms. Signal Process. Image Commun. **99**, 116505 (2021)
19. Pickup, L.: Machine learning in multi-frame image super-resolution. PhD thesis, Oxford University, UK (2007)
20. Laghrib, A., Hadri, A., Hakim, A., Raghay, S.: A new multiframe super-resolution based on nonlinear registration and a spatially weighted regularization. Inf. Sci. **493**, 34–56 (2019)
21. Zhu, Y.-N., Zhang, X.: Stochastic primal dual fixed point method for composite optimization. J. Sci. Comput. **84**(1), 16 (2020)
22. Zhu, Y.-N., Zhang, X.: A stochastic variance reduced primal dual fixed point method for linearly constrained separable optimization. SIAM J. Imag. Sci. **14**(3), 1326–1353 (2021)

Arabic Sign Language Classification Using CNN-LSTM Integration for Enhanced Gesture Recognition

Ihssane Bouhanou(✉) and Noureddine Aboutabit

Department of Computer System and Technology, ENSA, 25000 Khouribga, Morocco
Bouhanouihssane@gmail.com, n.aboutabit@usms.ma

Abstract. Arabic Sign Language (ArSL) recognition systems are essential tools for promoting inclusivity by facilitating communication between the Deaf community and individuals unfamiliar with sign language. By enabling effective communication, these systems can significantly enhance information accessibility and improve the daily experiences of Deaf individuals. In this project, we present an ArSL recognition approach using the KArSL database and an advanced model that combines Convolutional Neural Networks (CNN) and Long Short-Term Memory (LSTM) networks. Our model achieved over 85% accuracy, demonstrating its competitive performance and potential compared to other methodologies tested on this dataset. This high accuracy underscores the model's effectiveness in recognizing Arabic sign language signs and highlights the value of using deep learning techniques to bridge communication gaps within the Arabic-speaking Deaf community.

Keywords: Arabic sign language recognition (ArSLR) · continuous sign recognition · recognition · Convolutional Neural Networks (CNN) · LSTM

1 Introduction

Arabic Sign Language (ArSL) recognition systems play a crucial role in bridging communication gaps between the Deaf community and non-sign language users in Arabic-speaking regions. Through real-time translation capabilities, these systems enable Deaf individuals to access information and participate in daily interactions, fostering inclusivity and social engagement. Given that ArSL has been less studied compared to other sign languages, the development of precise recognition models for ArSL is vital for advancing accessibility and promoting equal opportunities for communication.

In this research, we focus on developing an advanced ArSL recognition system using the KArSL database, a valuable resource for Arabic Sign Language research. The system integrates Convolutional Neural Networks (CNNs) for effective spatial feature extraction with Long Short-Term Memory (LSTM) networks

A. Laghrib and A. Ghazdali (Eds.): ICMDS 2024, LNNS 1466, pp. 52–60, 2025.
https://doi.org/10.1007/978-3-031-94802-2_5

for capturing temporal relationships in gesture sequences. This combined approach enables the model to handle the complex dynamics of ArSL gestures, which involve both intricate hand movements and facial expressions.

Our work aims to contribute to the growing body of research in ArSL by developing a model capable of accurate recognition, potentially paving the way for more sophisticated applications. By leveraging CNN and LSTM architectures, we aim to enhance the accuracy and robustness of ArSL recognition systems. Additionally, this research seeks to advance sign language recognition technology, particularly for underrepresented languages, helping to ensure that Deaf individuals in the Arabic-speaking world have access to the same technological advancements as those available for more widely studied languages.

The paper is organized as follows: Sect. 2 briefly presents the literature on current sign language systems, and Sect. 3 provides an overview of the database used in the experiments. Section 4 describes the proposed recognition systems, and Sect. 5 discusses the outcomes of the experimental evaluation. Finally, Sect. 6 presents the conclusions.

2 Related Work

Sign language is a complex, multidimensional communication system that involves not only static hand gestures and finger spelling but also facial expressions, body posture, and movements. Each component contributes to conveying nuanced meanings and emotions, underscoring the unique grammar, syntax, semantics, and phonology of each sign language, including Arabic Sign Language (ArSL). This complexity demands specialized approaches tailored to each language's distinct linguistic traits, particularly for underrepresented languages such as ArSL, which require innovative methodologies to achieve accurate recognition. Each sign language, including Arabic Sign Language, is structured uniquely, with its own grammar, syntax, semantics, morphology, and phonology, reflecting its distinct linguistic characteristics. Recognizing these complexities calls for specialized approaches tailored to each language's unique traits, especially for languages like ArSL, which require further development and dedicated techniques for accurate recognition. Given the nuances within each gesture, current research strives to create sophisticated models that can decode these varied components effectively, ultimately advancing accessibility and understanding in sign language recognition systems.

In [1], the authors introduce KArSL, a comprehensive Arabic Sign Language dataset that supports the development of sign language recognition models tailored to Arabic-speaking deaf communities. The paper details the collection and annotation process, emphasizing the linguistic diversity within Arabic sign gestures. KArSL includes variations in regional dialects and gestures, providing an invaluable resource for advancing machine learning approaches in Arabic sign language recognition. This dataset has since facilitated research focused on improving model accuracy, robustness, and application relevance in realworld scenarios, making it a cornerstone for future developments in Arabic sign language accessibility.

In [2], Assaleh et al. developed a continuous Arabic Sign Language Recognition (ArSLR) system designed for a database of 40 sentences covering 80 commonly used words. Using spatio-temporal features extracted via the Discrete Cosine Transform (DCT), they employed an HMM classifier optimized by adjusting feature selection, state numbers, and Gaussian mixtures for improved performance. Building on this, Albelwi and Alginahi [16] advanced their initial alphabet recognition into a continuous, real-time ArSLR system. They used Fourier descriptors to model hand contours and achieved 90.6% accuracy with a KNN classifier on a limited database.

In [3], Tolba et al. implemented a continuous Arabic Sign Language Recognition (ArSLR) system using a Pulse-Coupled Neural Network (PCNN) and a graph matching technique. The system targets sentences comprising three to four words, segmenting the signs into elemental components and static postures before applying graph matching for pattern recognition. This method achieved an accuracy exceeding 70% on 30 continuous sentences containing a total of 100 gestures, demonstrating the potential of PCNN and graph-based methods for handling sequences in sign language data while maintaining robustness and flexibility across various sentence structures.

In the study conducted by Mohandes and Deriche [4], CyberGloves and hand-tracking devices were utilized for Arabic Sign Language recognition. The system used 22 signals from the glove's sensors and 6 from the tracking device for each hand. The dataset included 100 signs performed by a single signer, and the results showed accuracy rates of 84.7% for the tracking device and 91.38% for the glove-based system. The authors found that a traditional feature-based approach achieved a maximum accuracy of 96.2%, which was improved to 98.1% with decision-level fusion.

In contrast to Arabic Sign Language [1], numerous databases are available for non-Arabic sign languages, each contributing to the advancement of sign language recognition systems. For example, the UCIMachine Learning repository offers a dataset for Australian Sign Language, which is based on glove sensors and contains 95 signs performed by a single signer. Similarly, Barczak et al. [5] created a dataset for American Sign Language (ASL) with 36 static signs. Pugeault and Bowden also developed a database of ASL alphabets, collected using the Microsoft Kinect sensor. Moreover, the RWTH-BOSTON series, consisting of video-based databases (RWTH-BOSTON-50, RWTH-BOSTON-104, and RWTH-BOSTON-400), provides a broader dataset with a total of 843 sentences performed by five signers in the RWTH-BOSTON-400 dataset [6]. The SIGNUM database focuses on German Sign Language, featuring 780 sentences and 450 signs performed by 25 signers. Additionally, the ECHO database, available in Swedish Sign Language (SSL), Dutch Sign Language (NGT), and

British Sign Language (BSL), includes stories and interviews with signers, further expanding the variety of sign language datasets used for research and recognition system development [7]. These diverse datasets play a crucial role in improving the accuracy and applicability of sign language recognition technologies globally.

In [8], the authors explored various transformation methods for extracting and describing features from a series of sign frames accumulated into a single image. They evaluated the effectiveness of three transformation techniques—Fourier, Hartley, and Log-Gabor—on both the entire image and its segmented slices. Using different machine learning classifiers, they compared the performance of these transformations. Their findings indicated that the Hartley transformation achieved an impressive 98.8% accuracy for recognizing Arabic sign language with an SVM classifier.

3 Karsl Database

The KArSL database (KFUPM Arabic Sign Language) was developed to support research in Arabic Sign Language (ArSL) using data collected with Microsoft Kinect V2. This database includes 502 sign words that cover 11 thematic chapters commonly found in ArSL dictionaries: Letters, Numbers, Health, Common Verbs, Family, Characteristics, Directions and Places, Social Relationships, Household Terms, Religion, and Jobs and Professions. Three experienced male signers between the ages of 30 and 40 performed each sign word 50 times, resulting in a total of 75,300 sign samples (502 signs $\times$ 3 signers $\times$ 50 repetitions).

The signs in the KArSL database include both one-handed and two-handed signs—237 are one-handed, and 265 are two-handed. This mix introduces additional complexity, as certain body parts may become partially obscured during signing. Additionally, each sign consists of one to three distinct components, totaling 689 unique components across the database.

The KArSL database organizes signs hierarchically into three main categories: numbers, letters, and words. Each sign is assigned a unique five-digit identifier and is stored in three types of files: a color video file (MP4 format) and two binary files that contain depth and skeletal joint data. Each sample is saved with a standardized naming format.

4 Methodology

4.1 Convolutional Neural Network (CNN)

In this study, Convolutional Neural Networks (CNNs) were utilized due to their powerful capacity for feature extraction and pattern recognition in visual data [9,10]. CNNs operate through a sequence of convolutional and pooling layers, which enable a progressive learning process that captures spatial hierarchies [11,12]. Starting with basic elements like edges, CNNs gradually move toward identifying complex shapes, textures, and spatial relationships, making them well-suited for sign language recognition. This layered learning approach is essential in capturing the subtle, intricate movements and visual nuances in hand gestures within each frame of sign language video sequences, allowing CNNs to function as a robust feature extraction method tailored for accurate gesture interpretation and classification in sign language data.

4.2 Long Short-Term Memory (LSTM) Networks

Long Short-Term Memory (LSTM) networks are specialized Recurrent Neural Networks (RNNs) designed to preserve long-term dependencies within sequential data. Unlike traditional RNNs, LSTMs employ gating mechanisms that control information flow, selectively retaining or discarding data over time. This design makes them ideal for tasks requiring the understanding of sequential patterns, like interpreting the progression of gestures in sign language [8]. In this study, LSTMs contribute to capturing the temporal dynamics within sign language sequences, allowing the model to integrate information across multiple frames and enhance classification accuracy by retaining contextual nuances essential for precise recognition.

4.3 Model Architecture

The proposed model architecture integrates CNN and LSTM layers within a Keras framework to leverage both spatial and temporal feature extraction. The architecture begins with a TimeDistributed wrapper applied to the CNN layers, allowing individual processing of each video frame. The CNN segment includes several convolutional layers: an initial layer with 128 filters and a kernel size of (3, 3), followed by a second convolutional layer with 64 filters. Each convolutional layer utilizes ReLU activation and a stride of (1, 1), followed by max pooling operations to downsample the spatial dimensions.

Additional TimeDistributed convolutional layers further refine spatial features, and a BatchNormalization layer standardizes the output, promoting stability. Flattened CNN outputs are processed through a dropout layer with a rate of 0.2 to mitigate overfitting.

Next, the sequence is passed to a 32-unit LSTM layer, configured without return sequences and including a dropout rate of 0.2. The model concludes with two dense layers containing 64 and 32 units, respectively, both with ReLU activation. Finally, a softmax layer is applied for classification, outputting probabilities across 20 target classes, representing Arabic Sign Language signs (Fig. 1).

```
_________________________________________________________________
 Layer (type)                Output Shape              Param #
=================================================================
 time_distributed (TimeDistr  (None, 1, 248, 248, 128)  1280
 ibuted)

 time_distributed_1 (TimeDis  (None, 1, 246, 246, 64)  73792
 tributed)

 time_distributed_2 (TimeDis  (None, 1, 123, 123, 64)  0
 tributed)

 time_distributed_3 (TimeDis  (None, 1, 121, 121, 64)  36928
 tributed)

 time_distributed_4 (TimeDis  (None, 1, 119, 119, 32)  18464
 tributed)

 time_distributed_5 (TimeDis  (None, 1, 59, 59, 32)    0
 tributed)

 time_distributed_6 (TimeDis  (None, 1, 59, 59, 32)    128
 tributed)

 time_distributed_7 (TimeDis  (None, 1, 111392)        0
 tributed)

 dropout (Dropout)           (None, 1, 111392)         0

 lstm (LSTM)                 (None, 32)                14262400

 dense (Dense)               (None, 64)                2112

 dense_1 (Dense)             (None, 32)                2080

 dropout_1 (Dropout)         (None, 32)                0

 dense_2 (Dense)             (None, 20)                660

=================================================================
Total params: 14,397,844
Trainable params: 14,397,780
Non-trainable params: 64
```

Fig. 1. Structure of the Model.

5 Experimental Results

We present in this paper our sign recognition system using a subset of the KArSL database, which contains 20 unique signs performed by three different signers. The data was divided into 70% for training and 30% for testing. After tuning the algorithm's parameters, the model reached 85% accuracy on the test set, demonstrating promising performance.

To gain deeper insights, we analyzed the model's performance across individual classes, comparing true positives (TP), false positives (FP), and false negatives (FN) for each. The model performed particularly well in specific classes, notably class 2 (Precision: 1.0, F1: 0.933), class 3 (Precision and Recall: 0.917), and class 17 (Precision: 0.955, F1: 0.913), demonstrating a strong ability to recognize these signs accurately with minimal misclassification. This suggests that the model has successfully identified distinguishing features for these classes, leading to high recall and precision rates.

However, certain classes experienced more frequent misclassifications, notably class 5 (Precision: 0.676, F1: 0.793), class 11 (Precision: 0.667, F1: 0.772), and class 19 (Precision: 0.686, F1: 0.814). These lower precision scores imply that these classes might be commonly mistaken for other signs, likely due to visual similarities or overlaps with neighboring classes. This trend indicates a challenge in differentiation, potentially arising from insufficient inter-class variance or a lack of distinct features in the model's learned representations.

Lower-numbered classes, such as 0, 1, and 9, also showed scattered misclassifications. For example, class 9 had a relatively lower recall (0.708) and F1-score (0.773), suggesting that these signs may not be as easily distinguishable due to subtle differences or a limited dataset representation that hinders effective learning.

Furthermore, classes 5, 8, and 12 exhibited consistent misclassification patterns, indicating that additional refinement in feature extraction may be necessary to improve model performance. Enhanced methods could help increase separability for signs with visually similar gestures, reducing FP and FN rates.

Our analysis suggests several areas for enhancement, including the application of data augmentation to improve class balance and the refinement of model architecture for greater discriminative power. These adjustments may help reduce misclassification rates and enhance the model's ability to capture subtle differences across all sign classes. Overall, achieving a correct prediction rate of approximately 85% on the test set represents a significant step forward in accurate sign recognition, with room for further optimization to boost performance (Table 1).

Table 1. Detailed Performance Metrics for Each Sign Class

Class ID	True Positive	False Positive	False Negative	Recall (%)	Precision (%)	F1-Score (%)
0	23	3	1	95.83	88.46	91.84
1	22	6	2	91.67	78.57	84.62
2	21	0	3	87.50	100.00	93.33
3	22	2	2	91.67	91.67	91.67
4	20	2	4	83.33	90.91	86.96
5	23	11	1	95.83	67.65	79.31
6	19	0	5	79.17	100.00	88.37
7	19	0	5	79.17	100.00	88.37
8	23	6	1	95.83	79.31	86.79
9	17	3	7	70.83	85.00	77.27
10	17	1	7	70.83	94.44	80.95
11	22	11	2	91.67	66.67	76.19
12	21	1	3	87.50	95.45	91.30
13	23	5	1	95.83	82.14	88.46
14	18	1	6	75.00	94.74	83.72
15	20	1	4	83.33	95.24	88.89
16	19	1	5	79.17	95.00	86.36
17	21	1	3	87.50	95.45	91.30
18	19	1	5	79.17	95.00	86.36
19	24	11	0	100.00	68.57	81.36

6 Conclusion

This study presents an effective method for Arabic Sign Language Recognition (ArSLR) by integrating CNN and LSTM models, achieving an accuracy of over 85%. The approach leverages the strengths of convolutional neural networks (CNNs) to extract meaningful spatial features from sign language videos and utilizes long short-term memory (LSTM) networks to analyze the temporal sequences inherent in signing gestures. This combination allows the model to effectively capture the complex and dynamic nature of Arabic sign language, which involves both hand movements and facial expressions. The results emphasize the significant potential of deep learning techniques to improve ArSL accessibility, offering valuable tools to bridge communication gaps for Arabic-speaking Deaf communities. Furthermore, the study highlights the importance of technology-driven solutions in fostering more inclusive communication environments, empowering individuals with hearing disabilities to engage more effectively in social, educational, and professional settings. Future work will aim to refine the model's performance by incorporating advanced architectures and optimizing training techniques, expanding the dataset to include a wider range of signers and signing styles to improve generalization. Additionally, the study will explore real-world deployment scenarios to address practical challenges, such as recognizing signs in diverse lighting conditions and handling variations in signing speed, with the ultimate goal of promoting more inclusive communication solutions and contributing to the broader accessibility landscape.

Acknowledgment. We like to acknowledge the support provided by to the authors of the KArSL: Arabic Sign Language Database—Ala Addin I. Sidig, Hamzah Luqman, Sabri Mahmoud, and Mohamed Mohandes—for generously sharing their invaluable dataset and providing critical data to advance Arabic Sign Language recognition systems.

References

1. Sidig, A.A.I., Luqman, H., Mahmoud, S., Mohandes, M.: KArSL: Arabic Sign Language Database, King Fahd University of Petroleum and Minerals, Saudi Arabia (2022)
2. Assaleh, K., Shanableh, T., Fanaswala, M., Amin, F., Bajaj, H.: Continuous Arabic sign language recognition in user dependent mode (2010)
3. Elons, A.S., Abull-Ela, M., Tolba, M.F.: A proposed PCNN features quality optimization technique for poseinvariant 3D Arabic sign language recognition (2013)
4. Mohandes, M., Deriche, M.: Arabic sign language recognition by decisions fusion using Dempster-Shafer theory of evidence. In: Communications and IT Applications Conference (2013)
5. Barczak, A.L.C., Reyes, N.H., Abastillas, M., Piccio, A., Susnjak, T.: A new 2D static hand gesture colour image dataset for ASL gestures (2011)
6. Dreuw, P., Rybach, D., Deselaers, T., Zahedi, M., Ney, H.: Speech recognition techniques for a sign language recognition system (2007)
7. Crasborn, O.A., et al.: Sharing sign language data online: experiences from the ECHO project. Int. J. Corpus Linguist. **12**(4), 535–562 (2007)
8. Sidig, A.A.I., Luqman, H., Mahmoud, S.A.: Transform-based Arabic sign language recognition, King Fahd University of Petroleum and Minerals, Saudi Arabia (2017)
9. Nachaoui, A., Laghrib, A., Hakim, M.: Mathematical control and numerical applications. Springer (2021)
10. Nachaoui, M., Afraites, L., Hadri, A., Laghrib, A.: A non-convex non-smooth bi-level parameter learning for impulse and gaussian noise mixture removing. Commun. Pure Appl. Anal. **21**(4), 1249–1291 (2022)
11. El Malki, I., El Hakoume, A., Laghrib, A., Hadri, A.: Improved image denoising via self-supervised weickert operator learning and plug-and-play learned primal dual. Neurocomputing **620**, 129267 (2025)
12. Ghazdali, A., Ourdou, A., Hakim, M., Laghrib, A., Mamouni, N., Metrane, A.: Robust approach for blind separation of noisy mixtures of independent and dependent sources. Appl. Comput. Harmon. Anal. **60**, 426–445 (2022)

Triangle Regularization for Pruning Aware Training of Neural Networks

Abdelfattah Toulaoui(✉), Mourad Jabrane, Hamza Khalfi, and Imad Hafidi

Laboratory of Process Engineering, Informatics, and Mathematics, Sultan Moulay Slimane University, Beni Mellal, Morocco
abdelfattah.toulaoui1@usms.ac.ma

Abstract. Neural networks have increasingly been adopted in practical applications, leading to a growing need for improving their computational efficiency and energy consumption. Pruning has emerged as a key technique in addressing this challenge. Traditional pruning methods typically involve training a dense neural network to completion, followed by the removal of less significant weights or neurons. However, this approach often results in suboptimal performance due to a mismatch between the training and inference phases. In this paper, we propose a pruning-aware training method based on regularization using the tent (triangular) function. Our method enables the neural network to jointly optimize for both sparsity and performance during training. We evaluate our approach on benchmark datasets, including MNIST, CIFAR-10, and CIFAR-100, demonstrating its effectiveness and superior performance compared to traditional pruning techniques.

Keywords: Deep Neural Networks · Regularization · Pruning · Efficient Inference · Convolutional Neural Networks

1 Introduction

Neural networks have achieved remarkable success across various domains, but this success often comes at the cost of increased model complexity and over-parameterization. As the size of neural networks grows, their computational and memory demands also increase, creating challenges for deployment in resource-constrained environments. The increasing use of big data in various domains highlights the need for scalable and efficient and highly parallel algorithms [1–3]. Consequently, there is significant interest in methods that reduce model complexity without sacrificing performance. Among these, sparsity-inducing regularization techniques have garnered attention due to their ability to prune redundant parameters and enforce structured sparsity in the network [4,5].

Early work on pruning introduced methods to remove unimportant connections in neural networks, reducing their size and computational requirements while maintaining competitive accuracy [6]. Since then, various regularization techniques, such as ℓ_1 and ℓ_2 penalties, dropout, and group sparsity, have been

A. Laghrib and A. Ghazdali (Eds.): ICMDS 2024, LNNS 1466, pp. 61–69, 2025.
https://doi.org/10.1007/978-3-031-94802-2_6

proposed to promote sparsity during training [7]. Despite their success, these methods often rely on predefined thresholds or heuristics to identify parameters for pruning, which can limit their general applicability across different tasks and architectures.

In this paper, we present a novel approach for inducing sparsity in neural networks using the triangle function as a regularization term. The triangle function provides a smooth yet sharp penalization for weights near zero, encouraging the network to prune insignificant weights automatically during training. Unlike traditional sparsity-inducing regularizers, the triangle function introduces a non-linearity that facilitates adaptive thresholding, eliminating the need for manual tuning of sparsity levels [8,9]. This leads to a more flexible and task-agnostic pruning approach.

Our proposed method extends the current literature on sparsity regularization by introducing a mathematically principled, easily implementable technique that can be seamlessly integrated into existing deep learning frameworks. We demonstrate the effectiveness of our method on benchmark tasks, showing consistent reductions in model size without compromising accuracy and often outperforming conventional pruning methods.

The remainder of this paper is structured as follows. Section 2 provides a formulation of the pruning problem. Section 3 reviews existing work on pruning and sparsity-inducing regularization techniques. Section 4 introduces the mathematical formulation of our method, and Sect. 5 presents experimental results demonstrating its effectiveness across various domains.

2 Problem Formulation

Let $\mathbf{W} \in \mathbb{R}^{m \times n}$ represent the weight matrix of a neural network layer, where m is the number of input units and n is the number of output units. The goal of unstructured pruning is to reduce the number of non-zero elements in $\mathbf{W}$, thereby inducing sparsity while maintaining the overall performance of the model.

Unstructured pruning is traditionally applied post hoc to a fully trained model, where weights deemed less significant are set to zero. While this results in a sparser model with reduced computational demands, it often requires additional fine-tuning to recover accuracy. Our objective is to achieve an optimal balance between model sparsity and accuracy, enabling the network to generalize effectively while reducing computational costs.

Given a neural network parameterized by $\mathbf{W}$ and a loss function $\mathcal{L}(\mathbf{W})$, the unstructured pruning objective can be formulated as:

$$\min_{\mathbf{W}} \mathcal{L}(\mathbf{W}) \quad \text{subject to} \quad ||\mathbf{W}||_0 \leq k, \tag{1}$$

where $||\mathbf{W}||_0$ denotes the ℓ_0 norm of $\mathbf{W}$, counting the number of non-zero elements, and k is the target number of non-zero elements after pruning. This constraint reduces the network's size, enabling efficient inference on resource-limited hardware.

Since ℓ_0 norm optimization is combinatorially hard, it is often relaxed to the ℓ_1 norm, leading to the following optimization problem:

$$\min_{\mathbf{W}} \mathcal{L}(\mathbf{W}) + \lambda \|\mathbf{W}\|_1, \tag{2}$$

where $\|\mathbf{W}\|_1$ represents the sum of the absolute values of the elements of $\mathbf{W}$, and λ is a regularization parameter controlling the trade-off between minimizing the loss function and encouraging sparsity. This relaxation allows the use of continuous optimization methods, making the problem more tractable.

In practice, unstructured pruning is often performed iteratively as follows:

1. Train the neural network to obtain an initial weight matrix $\mathbf{W}^{(0)}$.
2. Identify a fraction p of weights with the smallest magnitudes in $\mathbf{W}^{(t)}$ and prune them to obtain a pruned weight matrix $\mathbf{W}^{(t+1)}$.
3. Fine-tune the pruned network by minimizing $\mathcal{L}(\mathbf{W}^{(t+1)})$, allowing the remaining weights to adapt and recover performance.
4. Repeat steps 2 and 3 until the desired level of sparsity is achieved or performance degradation exceeds an acceptable threshold.

Balancing the trade-off between sparsity and generalization remains a key challenge in pruning. Over-pruning risks underfitting, while under-pruning may leave the model too large to realize meaningful efficiency gains.

3 Related Works

Sparsity-inducing techniques have long been pivotal in reducing the complexity and size of neural networks, aiming to preserve model performance while minimizing the number of parameters. One of the foundational approaches to sparsity is pruning, introduced in early works such as Optimal Brain Damage [10] and Optimal Brain Surgeon [11]. These methods leverage second-order derivatives of the loss function to identify and remove weights with minimal impact on performance. These early contributions laid the groundwork for modern pruning techniques, demonstrating the effectiveness of removing redundant connections in neural networks.

Contemporary pruning methods have evolved towards simpler and more computationally efficient approaches, such as magnitude-based pruning. This widely adopted method removes weights based on their absolute values, typically focusing on those with smaller magnitudes [12]. Magnitude pruning is favored for its simplicity and effectiveness across a wide range of tasks. Beyond unstructured pruning, more sophisticated structured pruning techniques have emerged, aiming to remove entire neurons, channels, or filters, thereby yielding significant reductions in both memory usage and computational overhead [13]. These structured approaches are particularly beneficial for hardware acceleration, as they result in smaller, dense models that can be easily deployed on standard hardware like CPUs and GPUs [14]. A notable example of structured pruning is filter pruning, which selectively removes entire filters in convolutional layers based on learned importance scores [15].

In addition to post hoc pruning, several regularization techniques have been proposed to promote sparsity during training. Among these, ℓ_1 regularization remains one of the most prominent methods. By applying an absolute value penalty to the weights, ℓ_1 regularization encourages small weights to shrink to zero, naturally inducing sparsity in the model [16]. Despite its effectiveness, ℓ_1 regularization typically produces unstructured sparsity, which poses challenges for hardware that cannot fully exploit the resulting sparse weight matrices [17]. To overcome this, structured regularization techniques such as group Lasso have been developed. These methods extend the ℓ_1 penalty to groups of parameters, encouraging the removal of entire blocks or filters, thereby facilitating more hardware-efficient implementations [18]. Other advanced methods, such as dropout and its variational variant, aim to indirectly induce sparsity by randomly dropping units during training, promoting a less dependency on any specific neurons [19].

Sparsity-inducing regularization is further enhanced by approaches beyond conventional pruning and regularization. Quantization, for example, compresses neural networks by reducing the precision of weights. Techniques such as 8-bit or even 4-bit weight representations have been proposed, allowing for the use of low-precision compute units and improved memory efficiency on modern hardware [20,21]. However, quantization can degraded the model's performance when applied post-training [22]. To mitigate this issue, quantization-aware training techniques have been developed, offering superior performance by incorporating quantization into the training process [23].

Another important line of research involves neural network compression through low-rank factorization. Methods such as Singular Value Decomposition (SVD) [24], CP decomposition [25], and Tucker decomposition [26] decompose large weight matrices into several smaller, low-rank matrices, leveraging redundancies in the network's parameters. Among these, SVD has been particularly effective for dimensionality reduction, achieving compression by representing weight matrices as a product of unitary matrices and a diagonal matrix containing the singular values [27].

4 Proposed Approach

4.1 Triangle Regularization Function

To promote sparsity in deep neural networks (DNNs), we propose using a function that drives smaller parameters toward zero while encouraging larger parameters to remain large. For this purpose, we utilize the triangle function applied to positive values.

The function is defined in the interval $[-b, a]$ as follows:

$$t(x) = 1 - \max\left(\frac{x}{a}, -\frac{x}{b}\right).$$

In our case, the parameter magnitudes $|p_i|$ satisfy $|p_i| \in [0, \max(|p|)]$. We define $\alpha(p) = (1-r) \cdot \max(|p|)$ and $\beta(p) = r \cdot \max(|p|)$. Using these definitions, the previous equation can be rewritten for $(|p_i| - \beta(p)) \in [-\beta(p), \alpha(p)]$ as:

$$\mathcal{R}(p_i) = 1 - \max\left(\frac{|p_i| - \beta(p)}{\alpha(p)}, -\frac{|p_i| - \beta(p)}{\beta(p)}\right),$$

or equivalently, as a piecewise function:

$$\mathcal{R}(p_i) = \begin{cases} 1 - \frac{|p_i| - \beta(p)}{\alpha(p)}, & \text{if } |p_i| \geq \beta(p), \\ 1 + \frac{|p_i| - \beta(p)}{\beta(p)}, & \text{if } |p_i| < \beta(p). \end{cases}$$

Figure 1 illustrates the proposed regularization function, which is an even function. The regularization of each weight depends solely on the weights within the same matrix, enabling independent optimization across layers.

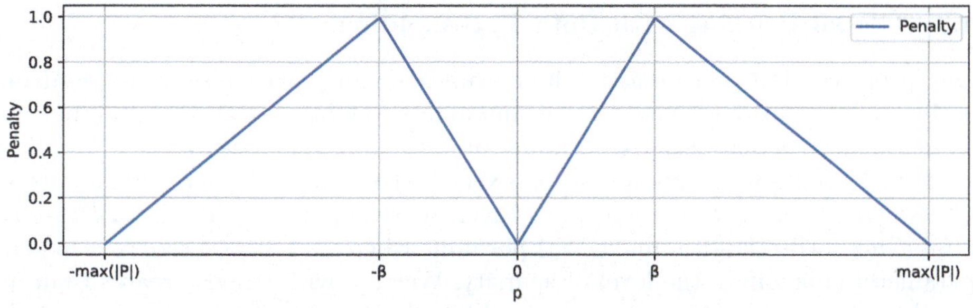

Fig. 1. The proposed triangle regularization function as a function of a parameter p, the maximum magnitude of the weight matrix it belongs to, and the hyperparameter β.

4.2 Derivative of the Function

The Adam optimizer [28] incorporates features such as momentum, which can potentially interfere with the application of regularization. To effectively use the proposed regularization function with Adam, we must compute the gradient of the function, and ensure that the regularization is applied as the final step of the optimization process.

First, we calculate the partial derivatives of $\alpha(p)$ and $\beta(p)$:

$$\frac{\partial \alpha(p)}{\partial p_i} = \begin{cases} \text{sign}(p_i) \cdot (1 - r), & \text{if } |p_i| = \max(|p|), \\ 0, & \text{if } |p_i| \neq \max(|p|), \end{cases}$$

and

$$\frac{\partial \beta(p)}{\partial p_i} = \begin{cases} \text{sign}(p_i) \cdot r, & \text{if } |p_i| = \max(|p|), \\ 0, & \text{if } |p_i| \neq \max(|p|). \end{cases}$$

Next, we derive the regularization function $\mathcal{R}(p_i)$:

$$\frac{\partial \mathcal{R}(p_i)}{\partial p_i} = \begin{cases} -\frac{\left(\text{sign}(p_i)-\frac{\partial \beta(p)}{\partial p_i}\right)\alpha(p)-(|p_i|-\beta(p))\frac{\partial \alpha(p)}{\partial p_i}}{\alpha^2(p)}, & \beta(p) \leq |p_i| \leq \max(|p|), \\ \frac{\left(\text{sign}(p_i)-\frac{\partial \beta(p)}{\partial p_i}\right)\beta(p)-(|p_i|-\beta(p))\frac{\partial \beta(p)}{\partial p_i}}{\beta^2(p)}, & 0 \leq |p_i| < \beta(p). \end{cases}$$

Upon simplification, we obtain:

$$\frac{\partial \mathcal{R}(p_i)}{\partial p_i} = \begin{cases} 0, & |p_i| \geq \max(|p|), \\ -\frac{\text{sign}(p_i)}{\alpha(p)}, & \beta(p) \leq |p_i| < \max(|p|), \\ \frac{\text{sign}(p_i)}{\beta(p)}, & 0 \leq |p_i| < \beta(p). \end{cases}$$

4.3 Global Pruning with Beta Thresholding

Our proposed Beta Pruning method prunes weights within a layer based on a threshold defined relative to the maximum weight magnitude, enabling an adaptive and context-sensitive pruning process across the network.

Beta Pruning retains weights that exceed a fraction β of the maximum absolute weight value within each unit. Given a weight matrix $\mathbf{W}$, Beta Pruning establishes a threshold $\beta \cdot \max(|\mathbf{W}|)$ per unit, where $\beta \in [0, 1]$ serves as a hyperparameter controlling the level of sparsity. Weights with values greater than or equal to this threshold are preserved, while all others are set to zero.

Formally, for the weight vector $\mathbf{w}_i$ of the i-th unit, a binary mask $\mathbf{m}_{ij}$ is computed as follows:

$$\mathbf{m}_{ij} = \begin{cases} 1 & \text{if } |w_{ij}| \geq \beta \cdot \max(|\mathbf{w}_i|), \\ 0 & \text{otherwise}, \end{cases} \tag{3}$$

where w_{ij} denotes the j-th weight in $\mathbf{w}_i$. This mask is then applied to retain only the significant weights.

The primary advantage of Beta Pruning lies in its adaptability, as the threshold is dynamically set based on each unit's maximum weight. This results in a unit-specific sparsity pattern that adjusts to the characteristics of the weight distribution. By tuning β, one can effectively control the level of sparsity: a lower β reduces pruning, while a higher β increases it, preserving only the most significant weights. This flexibility allows for a balanced trade-off between sparsity and performance, facilitating efficient inference while maintaining critical model representations necessary for accuracy in resource-constrained environments.

5 Experimental Evaluation

To test the effectiveness of our regularization term, we conduct experiments on multiple datasets and models. We compare the regularization on each network to a network trained without regularization.

For these experiments, we use a regularization weight of 10^{-3} and set $\beta(p) = 0.4 \times \max(|p|)$. All networks were trained for 200 epochs, and a learning rate that started from 10^{-5} for 5 epochs, 10^{-4} for 5 epochs, then 10^{-3}, and 10^{-2} for 40 epochs. The LR was then dropped to 10^-3 for 20 epochs and 10^{-4} for another 20 epochs. No regularization was applied for the first 20 epochs of training (Table 1).

Table 1. Pruning and fine-tuning results for various models. "None" indicates that no regularization was applied, while "TR" indicates our proposed method

Dataset	Model	Reg.	Acc.	#Params	Pruned Acc.	FT. Acc.
MNIST [29]	LeNet [30]	None	98.64	25.00%	12.52	97.80
		TR	98.79	25.00%	14.30	98.64
CIFAR-10 [31]	ResNet18 [32]	None	87.43	30.00%	14.09	86.16
		TR	87.33	30.00%	48.21	86.32
	ResNet34 [32]	None	89.35	30.00%	19.15	88.73
		TR	88.94	30.00%	32.21	88.81
	ResNet50 [32]	None	90.47	30.00%	10.42	88.82
		TR	89.64	30.00%	21.74	89.01
CIFAR-100 [31]	ResNet18 [32]	None	65.37	40.00%	9.26	63.99
		TR	65.32	40.00%	34.71	64.10
	ResNet34 [32]	None	68.00	40.00%	12.21	66.72
		TR	66.83	40.00%	39.96	66.50
	ResNet50 [32]	None	68.15	40.00%	4.86	66.68
		TR	67.59	40.00%	18.24	66.93

6 Conclusion

In this paper, we have introduced a novel approach for inducing sparsity in neural networks using the triangle function as a regularization term. Our method offers several advantages over traditional pruning and sparsity-inducing techniques, including the ability to automatically adapt pruning thresholds during training without the need for manual tuning. By leveraging the smooth yet sharp penalization properties of the triangle function, our approach effectively removes unimportant weights while preserving overall model performance.

Data Availability Statement. No novel data was used in this research, all the data referenced in publicly available.

Consent for Publication. All authors have consented to the publication of this manuscript.

References

1. Rochd, Y., Hafidi, I.: An efficient distributed frequent itemset mining algorithm based on spark for big data. Int. J. Intell. Eng. Syst. **12**(4), 367–377 (2019). https://doi.org/10.22266/ijies2019.0831.34 https://doi.org/10.22266/ijies2019.0831.34
2. Nasri, M., Hafidi, I., Metrane, A.: Multithreading parallel robust approach for the VRPTW with uncertain service and travel times. Symmetry **13**(1), 36 (2020). https://doi.org/10.3390/sym13010036
3. Rochd, Y., Hafidi, I., Ouartassi, B.: A review of scalable algorithms for frequent itemset mining for big data using hadoop and spark. In: Mizera-Pietraszko, J., Pichappan, P., Mohamed, L. (eds.) RTIS 2017. AISC, vol. 756, pp. 90–99. Springer, Cham (2019). https://doi.org/10.1007/978-3-319-91337-7_9
4. Liu, J., Tripathi, S., Kurup, U., Shah, M.: Pruning algorithms to accelerate convolutional neural networks for edge applications: a survey. CoRR abs/2005.04275 (2020). arXiv: 2005.04275
5. El Malki, I., El Hakoume, A., Laghrib, A., Hadri, A.: Improved image denoising via self-supervised weickert operator learning and plug-and-play learned primal dual. Neurocomputing **620**, 129267 (2025)
6. Sietsma, Dow: Neural net pruning-why and how. In: IEEE 1988 International Conference on Neural Networks, pp. 325–3331 (1988). https://doi.org/10.1109/ICNN.1988.23864
7. Yang, C., et al.: Structured pruning of convolutional neural networks via L1 regularization. IEEE Access **7**, 106385–106394 (2019). https://doi.org/10.1109/access.2019.2933032
8. El Hakoume, A., Laghrib, A., Afraites, L., Hadri, A.: A variational PDNet network using a learning reaction-diffusion equation. Expert Syst. Appl. **249**, 123605 (2024)
9. Limami, F.-E., Hadri, A., Afraites, L., Laghrib, A.: Tensor-guided learning for image denoising using anisotropic PDEs. Mach. Vis. Appl. **35**(3), 48 (2024)
10. Cun, Y.L., Denker, J.S., Solla, S.A.: Optimal brain damage, pp. 598–605. Morgan Kaufmann Publishers Inc., San Francisco (1990)
11. Hassibi, B., Stork, D.G.: Second order derivatives for network pruning: optimal brain surgeon. In: Proceedings of the 5th International Conference on Neural Information Processing Systems. NIPS 1992, pp. 164–171. Morgan Kaufmann Publishers Inc., San Francisco (1992)
12. Gupta, M., et al.: Is complexity required for neural network pruning? A case study on global magnitude pruning. In: 2024 IEEE Conference on Artificial Intelligence (CAI), pp. 747–754 (2024). https://doi.org/10.1109/CAI59869.2024.00144
13. He, Y., Xiao, L.: Structured pruning for deep convolutional neural networks: a survey. IEEE Trans. Pattern Anal. Mach. Intell. **46**(5), 2900–2919 (2024). https://doi.org/10.1109/TPAMI.2023.3334614
14. Liu, Z., Sun, M., Zhou, T., Huang, G., Darrell, T.: Rethinking the Value of Network Pruning (2019)
15. Li, J., Shao, H., Deng, X., Jiang, Y.: Efficient filter pruning: reducing model complexity through redundancy graph decomposition. Neurocomput. **599**(C) (2024). https://doi.org/10.1016/j.neucom.2024.128108
16. Kurtz, M., et al.: Inducing and exploiting activation sparsity for fast inference on deep neural networks. In: III, H.D., Singh, A. (eds.) Proceedings of the 37th International Conference on Machine Learning. Proceedings of Machine Learning Research, vol. 119, pp. 5533–5543. PMLR (2020). https://proceedings.mlr.press/v119/kurtz20a.html

17. Zhang, S., et al.: Cambricon-x: an accelerator for sparse neural networks. In: 2016 49th Annual IEEE/ACM International Symposium on Microarchitecture (MICRO), pp. 1–12 (2016). https://doi.org/10.1109/MICRO.2016.7783723
18. Scardapane, S., Comminiello, D., Hussain, A., Uncini, A.: Group sparse regularization for deep neural networks. Neurocomputing **241**, 81–89 (2017). https://doi.org/10.1016/j.neucom.2017.02.029
19. Park, S., Kwak, N.: Analysis on the dropout effect in convolutional neural networks. In: Lai, S.-H., Lepetit, V., Nishino, K., Sato, Y. (eds.) ACCV 2016. LNCS, vol. 10112, pp. 189–204. Springer, Cham (2017). https://doi.org/10.1007/978-3-319-54184-6_12
20. Liu, B., Li, F., Wang, X., Zhang, B., Yan, J.: Ternary weight networks. In: ICASSP 2023 - 2023 IEEE International Conference on Acoustics, Speech and Signal Processing (ICASSP). IEEE (2023). https://doi.org/10.1109/icassp49357.2023.10094626
21. Banner, R., Hubara, I., Hoffer, E., Soudry, D.: Scalable methods for 8-bit training of neural networks. In: Proceedings of the 32nd International Conference on Neural Information Processing Systems. NIPS 2018, pp. 5151–5159. Curran Associates Inc., Red Hook (2018)
22. Nahshan, Y., et al.: Loss aware post-training quantization. Mach. Learn. **110**(11–12), 3245–3262 (2021). https://doi.org/10.1007/s10994-021-06053-z
23. Hubara, I., Courbariaux, M., Soudry, D., El-Yaniv, R., Bengio, Y.: Quantized neural networks: training neural networks with low precision weights and activations. J. Mach. Learn. Res. **18**(187), 1–30 (2018)
24. Tian, N., Liu, Y., Wang, W., Meng, D.: Fast CNN inference by adaptive sparse matrix decomposition. In: 2021 International Joint Conference on Neural Networks (IJCNN). IEEE (2021). https://doi.org/10.1109/ijcnn52387.2021.9533546
25. Astrid, M., Lee, S.-I.: CP-decomposition with tensor power method for convolutional neural networks compression. In: 2017 IEEE International Conference on Big Data and Smart Computing (BigComp). IEEE (2017). https://doi.org/10.1109/bigcomp.2017.7881725
26. Liu, Y., Ng, M.K.: Deep neural network compression by tucker decomposition with nonlinear response. Knowl.-Based Syst. **241**, 108171 (2022). https://doi.org/10.1016/j.knosys.2022.108171
27. Stewart, G.W.: On the early history of the singular value decomposition. SIAM Rev. **35**, 551–566 (1993)
28. Kingma, D.P., Ba, J.: Adam: a method for stochastic optimization. CoRR abs/1412.6980 (2014)
29. LeCun, Y., Cortes, C., Burges, C.: MNIST handwritten digit database. ATT Labs, vol. 2 (2010). http://yann.lecun.com/exdb/mnist
30. Lecun, Y., Bottou, L., Bengio, Y., Haffner, P.: Gradient-based learning applied to document recognition. Proc. IEEE **86**(11), 2278–2324 (1998). https://doi.org/10.1109/5.726791
31. Krizhevsky, A.: Learning multiple layers of features from tiny images (2009). https://api.semanticscholar.org/CorpusID:18268744
32. He, K., Zhang, X., Ren, S., Sun, J.: Deep residual learning for image recognition. CoRR abs/1512.03385 (2015). arXiv:1512.03385

PrivFlow-NIDS: Privacy-Preserving Network Intrusion Detection Through Federated Learning and Normalizing Flows

Anass Abdelhamid El Alami[1](✉), Meryem Janati Idrissi[2], Mohamed Boussakssou[1], Mohamed Sebnat[3], and Abdelali Elmoufidi[4]

[1] Data4Earth, Sultan Moulay Slimane University, Beni Mellal, Morocco
anass-abdelhamid.el-alami@usms.ma

[2] College of Computing, Mohammed VI Polytechnic University, Benguerir, Morocco
Meryem.janati@um6p.ma

[3] TIAD, Sultan Moulay Slimane University, Beni Mellal, Morocco
m.sebnat@usms.ma

[4] LARPEG, National School of Commerce and Management Casablanca, Hassan 2 University, Casablanca, Morocco
a.elmoufidi@encgcasa.ma

Abstract. The growing volume and sophistication of cyberattacks have driven research into effective network intrusion detection systems (NIDS). Anomaly-based network intrusion detection approaches have become increasingly prominent due to their effectiveness in identifying unknown and sophisticated attacks. However, many existing methods rely on centralized architectures, raising significant privacy concerns due to the need for aggregating sensitive data in one location. Additionally, a common assumption in these methods is the availability of only normal data, which does not reflect real-world scenarios where limited attack samples are often available. In this paper, we propose, PrivFlow-NIDS, a semi-supervised, decentralized approach that combines Normalizing Flows (NFs) and Federated Learning (FL). The NF model is leveraged for data augmentation enriching scarce anomalous samples. By learning the underlying distribution of normal traffic, NF generates pseudo-attacks from the distribution's tails. Meanwhile, FL ensures privacy-preserving collaborative model training across distributed NIDS. We evaluate PrivFlow-NIDS using three well-known datasets, USTC-TFC2016, CIC-IDS2017, and UNSW-NB15. The results demonstrate that, with limited anomalous samples, PrivFlow-NIDS achieves high accuracy, detection rate, and macro F1-score, coupled with a low false alarm rate (99.91%, 87.59%, and 98.28% F1-score for USTC-TFC2016, CIC-IDS2017, and UNSW-NB15 respectively). Moreover, PrivFlow-NIDS performs on par with centralized settings, with differences of less than 0.1% in most cases. These findings highlight that integrating NFs and FL within IDS offers a promising, privacy-preserving solution for large-scale network intrusion detection.

M. Janati Idrissi, M. Boussakssou, M. Sebnat, and A. Elmoufidi—Contributed equally to this work.

A. Laghrib and A. Ghazdali (Eds.): ICMDS 2024, LNNS 1466, pp. 70–81, 2025.
https://doi.org/10.1007/978-3-031-94802-2_7

Keywords: Network Intrusion Detection · Federated Learning · Normalizing Flows · Anomaly Detection

1 Introduction

As cyberattacks grow in both complexity and frequency, Network Intrusion Detection Systems (NIDS) [1] have become vital in protecting sensitive data and networks. NIDS are tools or software designed to monitor network traffic, detect potential malicious activities, and alert security teams to suspicious behavior. Traditionally, NIDS are categorized into two types: signature-based (SNIDS) and anomaly-based (ANIDS) [2]. SNIDS work by comparing network traffic to a database of known attack signatures, making them effective at identifying familiar threats. However, they struggle to detect novel or previously unknown attacks. In contrast, ANIDS operate by learning the normal behavior of a system and flagging any deviations as potential anomalies. ANIDS have shown promise in identifying unknown threats, including zero-day attacks, but they face challenges in establishing a reliable baseline for normal behavior, often resulting in higher false alarm rates. Additionally, the computational demands of ANIDS can impact real-time detection capabilities. Despite these challenges, NIDS continue to play a crucial role in cybersecurity by defending against a wide array of threats.

Numerous machine learning (ML)-based approaches have been proposed for network intrusion detection systems (NIDS) [3], leveraging algorithms that classify network traffic to detect anomalies and potential threats. Generative models, particularly, have emerged as powerful tools for anomaly-based intrusion detection systems by enabling the detection of novel and sophisticated cyber threats [4,5]. Unlike traditional signature-based methods, which rely on pre-existing knowledge of attack patterns, generative models learn the distribution of normal network behavior, allowing them to identify deviations as potential threats. These models, including Variational Autoencoders (VAEs) [6] and Generative Adversarial Networks (GANs) [7], are well-suited for this task as they can capture complex patterns within data. Additionally, generative models have another advantage in creating synthetic data that mimics anomalous activities. By generating realistic synthetic samples, they enhance model training, enabling IDS to better detect rare or unseen attacks. This capability not only strengthens the system's adaptability but also improves its resilience against the constantly evolving landscape of cyber threats.

However, many of the proposed approaches operate under the assumption that only normal (benign) data is available for training [8–10]. However, in practice, limited samples of known attacks are often available, and their inclusion could significantly enhance detection capabilities. On the other hand, most of these methods are implemented in a centralized setting [11,12], which overlooks critical data privacy concerns inherent in real-world applications. Centralized models require that data from various sources be aggregated in a single location, potentially exposing sensitive information and increasing vulnerability to data breaches.

In this paper, we propose a semi-supervised federated learning method for NIDS, leveraging the capabilities of Normalizing Flows (NFs) [13], a class of generative models that can learn complex high-dimensional distributions and enable efficient anomaly detection. Our approach combines the data-efficient, anomaly-detection strengths of NFs with the privacy-preserving benefits of federated learning. The semi-supervised aspect of our method enables it to utilize both available attack samples and a larger pool of normal data, making it well-suited to real-world scenarios where labeled intrusion data is sparse. By integrating NFs, our method is capable of capturing complex data distributions, allowing the model to detect subtle deviations indicative of both known and emerging threats. Finally, federated learning, allows multiple NIDS entities to collaboratively train a model without the need to share raw data. The main contributions of this paper are as follows:

1. We introduce PrivFlow-NIDS, a semi-supervised approach for anomaly-based intrusion detection incorporating both benign and limited anomalous data.
2. We build the proposed method by leveraging NFs for data augmentation of anomalous data and employ federated learning to preserve data privacy within a decentralized NIDS.
3. We conduct experiments using a clean version of three well-known NIDS datasets, USTC-TFC2016, CIC-IDS2017, and UNSW-NB15 from which more reliable features were extracted using an updated version of CICFlowMeter. The results indicate that the proposed method demonstrates outstanding performance across various metrics.

The rest of this paper is organized as follows: In Sect. 2, related work on normalizing flows and federated learning in IDS is discussed. Section 3 presents the proposed method and discusses its key components. Section 4 delves into the implementation details and discusses the evaluation results. Finally, the paper is concluded in Sect. 5.

2 Related Work

2.1 Normalizing Flows in IDS

In the context of network intrusion detection, NFs present an opportunity to model normal network behavior with high fidelity while capturing subtle anomalies that might indicate potential threats. By learning the underlying distribution of normal network traffic patterns, NFs can identify suspicious activities that deviate from this learned distribution. Furthermore, their ability to generate synthetic samples from the learned distribution opens new possibilities for augmenting training data and improving detection capabilities, particularly in scenarios where attack samples are scarce. In this respect, Zhangxuan Dang et al. [14] proposed a method that involves a three-stage framework for anomaly detection using only normal traffic. It employs a reconstruction method to learn deep representations of normal samples, normalizes these representations using a bidirectional flow module, and generates pseudo anomaly samples by introducing noise to these normalized features. The final stage involves training a classifier to differentiate between normal and pseudo anomaly samples. The results demonstrate that this method achieves state-of-the-art performance on common benchmarking datasets for anomaly network traffic detection. However, the limitations include the need for high-quality normal data for training and the potential challenges in generalizing the generated pseudo anomalies to real-world anomalies [15]. Brian Matejek et al. [16] proposed a novel deep learning framework for network intrusion detection that includes a model-agnostic safeguard to quantify uncertainty in model decisions. This safeguard, which uses normalizing flows to learn class-conditional representations of internal features, enhances the model's robustness against zero-day attacks and concept drift. The authors demonstrated the effectiveness of their approach on the CIC-IDS-2017 and UNSW-NB15 datasets, achieving good results in detecting novel inputs. However, the study is limited by the use of only two datasets and the potential variability of the safeguard's effectiveness depending on the specific deep learning architecture used [17–19]. The authors of [20] proposed NF-NIDS, a novel approach that leverages normalizing flows for network intrusion detection in scenarios with limited attack samples. Their method employs two flow-based models; Inverse Autoregressive Flow (IAF) and Neural Spline Flows (NSF) to learn the underlying distribution of normal network traffic and generate pseudo-attacks from the distribution tails. To address the challenge of scarce attack data, they developed a semi-supervised learning scheme that combines normal traffic data with both available real attack samples and generated pseudo-anomalies to train a classifier. The approach was evaluated on three prominent datasets (USTC-TFC2016, CIC-IDS2017, and CSE-CIC-IDS2018) using only 0.02–0.08% of the attack samples for training. NF-NIDS achieved impressive results with accuracies of 98.71%, 94.86%, and 98.20% respectively across the datasets while maintaining negligible false discovery rates. Their work demonstrated that normalizing flows can effectively model network behavior distributions and generate synthetic attacks to enhance intrusion detection performance when labeled attack data is limited.

2.2 Federated Learning for Anomaly-Based IDS

Federated Learning (FL) has emerged as a promising paradigm for collaborative model training in network intrusion detection systems, particularly addressing privacy concerns and data silos in distributed environments. In FL-based NIDS, multiple participants can jointly train intrusion detection models without sharing their raw network traffic data, thereby preserving data privacy while benefiting from collective knowledge. This approach has gained considerable attention in recent years, especially for IoT and industrial environments where data privacy and resource constraints are critical concerns. In exploring federated anomaly-based NIDS, Ayed et al. [21] implemented a CNN model with federated averaging on the CICIDS2017 dataset, testing four client selection strategies: all clients, random selection, attack-based selection, and operating system-based selection.

While their federated model maintained an accuracy of 93%, slightly lower than centralized learning (98.8%), it effectively preserved data privacy by retaining sensitive data locally. The approach demonstrated robustness even when clients intermittently disconnected, underscoring its suitability for real-world environments. Building on privacy-preserving federated frameworks, Janati Idrissi et al. [22] developed Fed-ANIDS, a NIDS that employs three types of autoencoders (simple, variational, and adversarial) and the FedProx algorithm for federated learning. Tested on multiple datasets (USTC-TFC2016, CIC-IDS2017, and CSE-CIC-IDS2018), Fed-ANIDS achieved a maximum F1-score of 99.94% and an accuracy of 99.95%, outperforming GAN-based models with a low false detection rate. FedProx proved more effective than FedAvg, particularly in handling heterogeneous data, although the increased complexity may present deployment challenges. Jin et al. [23] addressed the problem of catastrophic forgetting in incremental intrusion detection with their FL-IIDS model. The system incorporates a dynamic example memory and specialized loss functions (class gradient balance and sample label smoothing) to preserve previous knowledge in distributed environments. Evaluated on the UNSW-NB15 and CICIDS2018 datasets, FL-IIDS demonstrated high accuracy, with performance degradation of only 0.374% to 0.980% on new attack classes. The approach, particularly effective with the CNN-GRU architecture, shows promise for intrusion detection, though further validation in diverse environments is warranted. AbdulRahman et al. [24] proposed a federated learning framework for intrusion detection in IoT networks, utilizing local model training and model aggregation to enhance privacy and detection accuracy. The system, tested on the NSL-KDD dataset, achieved performance comparable to centralized models and minimized communication overhead. However, the study's scope is limited by its reliance on a single dataset, and it does not address device heterogeneity or the potential for attacks targeting federated processes. Finally, Li et al. [25] presented FLEAM, an architecture combining FL and fog/edge computing to detect and mitigate DDoS attacks in Industrial IoT. Utilizing an iterative model averaging (IMA) gated recurrent unit (GRU) protocol, FLEAM achieved a 98% detection accuracy and a 72% reduction in mitigation response time, alongside a 2.7-fold increase in attacker costs. Although promising, the evaluation was limited to simulated DDoS attacks, and further research is needed to assess FLEAM's effectiveness against diverse real-world threats.

3 Our Method

We introduce PrivFlow-NIDS, a novel framework that employs Network Functions (NFs) to enhance Network Intrusion Detection Systems (NIDS) within a federated learning paradigm. By applying NFs to network traffic data across multiple participants while safeguarding data privacy, we can accurately model the underlying probability distribution of normal network behavior and derive a probability density function to quantify the likelihood of observed benign events. Figure 1 provides a comprehensive overview of PrivFlow-NIDS, which comprises three primary phases: local data preprocessing, federated model training, and the detection phase. A detailed explanation of each step is provided in the subsequent sections.

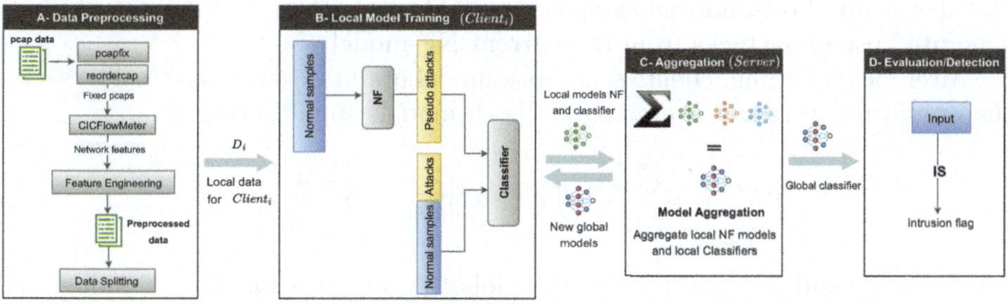

Fig. 1. PrivFlow-NIDS general architecture which consists of four main phases, data preprocessing, local model training, aggregation, and a detection phase.

3.1 Data Preprocessing

The preprocessing of network traffic data follows the approach in [26,27], starting with pcapfix for corrupted pcap files and reordercap for temporal sequencing. Feature extraction using CICFlowMeter generates 87 network traffic characteristics in CSV format. We remove flow identification features (ID, IP addresses, ports, timestamps) to prevent bias, and implement Feature Conversion to transform categorical variables into numerical representations. Feature Normalization is applied using min-max scaling: $x' = \frac{x - x_{min}}{x_{max} - x_{min}} \in \mathbb{R}^{82}$, where x_{min} and x_{max} are derived from each client's local training dataset.

3.2 Federated Training

This phase involves the simultaneous training of two models: a normalizing flow (NF) model and a classifier. The federation server initializes both models and coordinates their concurrent training while ensuring data privacy.

At the beginning of each communication round t, the server selects a fraction f of available clients, where $f \in (0, 1]$. For a total of N clients, the number of selected clients is computed as $m = \max(f \cdot N, 1)$. Selected clients receive both the current global NF model and classifier, then proceed with the following steps:

Local NF Training: Each selected client k trains the NF model using their normal network traffic data, minimizing the negative log-likelihood:

$$\mathcal{L}local^{NF} = -\frac{1}{|D_k|} \sum x \in D_k \log p_X(x)$$

Pseudo-attack Generation: Immediately following the local NF update, each client utilizes their updated NF model to generate synthetic attack samples by sampling from the tails of the learned distribution.

Classifier Training: In parallel with the NF training, each client updates the classifier using: Local normal samples, available real attack instances, and newly generated pseudo-attacks from the current NF model.

After local training, clients send both model updates to the server. The server then performs weighted averaging for both models simultaneously:

$$\theta_{global}^{NF} = \sum_{k=1}^{m} \frac{n_k}{n} \theta_k^{NF}, \quad \theta_{global}^{clf} = \sum_{k=1}^{m} \frac{n_k}{n} \theta_k^{clf}$$

where θ_{global}^{NF} and θ_{global}^{clf} represent the global parameters for the NF model and classifier respectively, n_k is the size of client k's local dataset, and n is the total number of samples across selected clients.

The server then broadcasts the updated global models to all clients for the next round. This concurrent training approach allows the classifier to continuously adapt to the improving quality of synthetic attacks generated by the evolving NF model, while the NF model benefits from the classifier's feedback through the training process.

Detection Phase. Once the convergence of both the federated NF model and classifier training phases is achieved, the detection phase begins. Each client utilizes their local copy of the global classifier to classify incoming network traffic as either normal or attack. First, an Intrusion Score (IS) is computed for the input data, which, in our case, are the probabilities output by the classifier. If the IS exceeds a certain threshold, the traffic is classified as an attack.

$$x_{new} = \begin{cases} Normal, & \text{if IS} \leq \text{thr} \\ Anomaly, & \text{otherwise} \end{cases} \tag{1}$$

This federated approach enables organizations to collaboratively build robust intrusion detection models while keeping their sensitive network data local and private. The resulting model benefits from the diversity of network environments and attack patterns across participating organizations while maintaining strict data privacy requirements.

4 Results and Discussion

4.1 Implementation Details and Valuation Metrics

The NSF model consists of 10 bijective neural networks, each with two hidden layers, and a three-layer fully connected classifier. We experiment with 100 attack samples (0.04%, 0.03%, 0.2%) and 1,000 attack samples (0.4%, 0.3%, 2%) of the USTC-TFC2016, CIC-IDS2017, and UNSW-NB15 datasets respectively. The threshold was empirically selected from 0.80, 0.81, ..., 0.99 to maximize the testing F1-score.

For centralized training, the NSF model uses 10 epochs, batch size 100, and learning rate 2E-04, while the classifier uses 3 epochs, batch size 100, and learning rate 0.001 with Adam optimizer.

In federated learning, data is equally distributed among K=10 clients, with 50% randomly selected per round. Selected clients train local NSF models for 5 epochs and classifiers for 3 epochs, over 10 communication rounds, followed by server aggregation.

Performance is evaluated using macro F1-score (chosen for imbalanced datasets), Accuracy, False Discovery Rate (FDR), and detection rate, with FDR measuring the system's false positive rate.

4.2 Performance Analysis of PrivFlow-NIDS

Our experimental evaluation assessed the performance of the proposed normalizing flows-based Network Intrusion Detection System in both centralized and federated (PrivFlow-NIDS) settings. We conducted experiments on three distinct datasets (USTC-TFC2016, CIC-IDS2017, and UNSW-NB15) with varying numbers of attack instances (100 and 1000) to evaluate the system's robustness and scalability.

Table 1. Results of using 100 and 1000 anomalies on USTC-TFC2016

Metrics	100 anomalies		1000 anomalies	
	Centralized	PrivFlow-NIDS	Centralized	PrivFlow-NIDS
F1-score (%)	0.997	0.9969	99.91	0.9991
FDR (%)	0.0052	0.0053	0.1498	0.0016
Accuracy (%)	99.7	99.7	99.92	99.91
DR (%)	1.0	1.0	1.0	1.0

Table 2. Results of using 100 and 1000 anomalies on CIC-IDS2017

Metrics	100 anomalies		1000 anomalies	
	Centralized	PrivFlow-NIDS	Centralized	PrivFlow-NIDS
F1-score (%)	98.12	71.85	96.14	98.28
FDR (%)	0.9871	0.0466	0.8376	0.8085
Accuracy (%)	98.12	73.14	96.14	98.28
DR (%)	97.41	49.06	93.47	97.54

Table 3. Results of using 100 and 1000 anomalies on UNSW-NB15

Metrics	100 anomalies		1000 anomalies	
	Centralized	PrivFlow-NIDS	Centralized	PrivFlow-NIDS
F1-score (%)	87.55	87.58	87.55	87.59
FDR (%)	0.004	0.0073	0.0026	0.0071
Accuracy (%)	97.96	97.97	97.96	97.97
DR (%)	97.9	97.91	97.9	97.91

4.2.1 USTC-TFC2016 Dataset Results

For the USTC-TFC2016 dataset, our approach demonstrated exceptional performance across both scenarios. As shown in Table 1, with 100 anomalies, the centralized approach achieved an accuracy of 99.7% and an F1-score of 0.997, while PrivFlow-NIDS showed identical accuracy (99.7%) and a comparable F1-score of 0.9969. Both approaches maintained a perfect Detection Rate (DR) of 1.0, with minimal False Detection Rates (FDR). When scaling to 1000 anomalies, both approaches showed slight performance improvements, with accuracy reaching 99.92% and 99.91% for centralized and federated settings respectively.

4.2.2 CIC-IDS2017 Dataset Results

The evaluation on the CIC-IDS2017 dataset revealed interesting dynamics between centralized and federated approaches, as shown in Table 2. For the 100 anomalies scenario, the centralized setting achieved an accuracy of 98.12% and an F1-score of 98.12%, while PrivFlow-NIDS showed lower performance with 73.14% accuracy and 71.85% F1-score. However, when scaling to 1000 anomalies, PrivFlow-NIDS demonstrated remarkable improvement, achieving 98.28% accuracy and F1-score, slightly outperforming the centralized approach (96.14%). The detection rate also showed substantial improvement, increasing from 49.06% with 100 anomalies to 97.54% with 1000 anomalies in the federated setting.

4.2.3 UNSW-NB15 Dataset Results

The UNSW-NB15 dataset evaluation revealed consistent performance across both attack scenarios (Table 3). With 100 anomalies, the centralized approach achieved an accuracy of 97.96% and an F1-score of 87.55%, while PrivFlow-NIDS showed slightly better results with 97.97% accuracy and 87.58% F1-score. The performance remained stable when increasing to 1000 anomalies, with both approaches maintaining high detection rates and extremely low false detection rates.

4.3 Discussion

Several key findings emerge from our experimental results:

1. **Privacy-Preserving Performance:** PrivFlow-NIDS consistently achieved performance metrics comparable to, and in some cases slightly better than, the centralized approach, validating our hypothesis that federated learning can maintain detection effectiveness while providing enhanced privacy guarantees.
2. **Scalability:** The system demonstrated varying degrees of scalability across datasets. While USTC-TFC2016 and UNSW-NB15 showed consistent performance, CIC-IDS2017 revealed significant improvements with increased attack samples in the federated setting.
3. **Dataset Adaptability:** The system showed strong adaptability across different datasets, achieving high accuracy rates on USTC-TFC2016 (>99%), CIC-IDS2017 (>98% with 1000 anomalies), and UNSW-NB15 (>97%).
4. **Sample Size Sensitivity:** The CIC-IDS2017 results demonstrate that our federated approach can be particularly sensitive to the number of attack samples in certain network environments, showing significant performance improvements when scaling from 100 to 1000 anomalies.
5. **False Detection Control:** Both approaches maintained remarkably low false detection rates across most scenarios, with particularly impressive results on UNSW-NB15 (order of 10^{-3}), indicating high precision in attack detection.

These results demonstrate that our proposed approach successfully achieves the dual objectives of maintaining high detection performance while enabling privacy-preserving distributed learning. The consistent performance across different datasets and attack scenarios, along with the notable improvements in larger sample sizes, highlights the robustness and practical applicability of our approach.

5 Conclusion

This study introduces PrivFlow-NIDS, a novel framework that integrates normalizing flows with federated learning for enhanced network intrusion detection. The proposed architecture employs a semi-supervised learning paradigm, leveraging both existing attack signatures from established datasets and synthetically generated samples through our normalizing flow model, which serves as a data augmentation mechanism. Through comprehensive empirical evaluation using three distinct datasets (USTC-TFC2016, CIC-IDS2017, and UNSW-NB15), we demonstrate the efficacy and practical viability of our proposed methodology.

The experimental results show that PrivFlow-NIDS achieves performance comparable to centralized approaches while offering enhanced privacy benefits. On the USTC-TFC2016 dataset, our federated approach achieved exceptional accuracy (99.7% and 99.91% for 100 and 1000 anomalies respectively) with perfect detection rates (DR = 1.0). For the CIC-IDS2017 dataset, our approach demonstrated significant improvement when scaling to larger attack samples, achieving 98.28% accuracy with 1000 anomalies, outperforming the centralized approach. Similarly, on the UNSW-NB15 dataset, PrivFlow-NIDS maintained robust performance with 97.97% accuracy and detection rates of 0.9791 across both scenarios. These results demonstrate that the privacy-preserving benefits of federated learning can be achieved without significant compromise in detection capabilities, and in some cases, can even enhance performance.

The successful implementation and evaluation of PrivFlow-NIDS opens several promising directions for future research:

- Investigation of the system's performance under different federation strategies and participant configurations
- Exploration of alternative normalizing flow architectures, including Inverse Autoregressive Flows (IAF), GLOW, Continuous Normalizing Flows (CNFs) and other advanced variants, to potentially enhance the system's detection capabilities and efficiency

References

1. Ahmad, Z., Shahid Khan, A., Wai Shiang, C., Abdullah, J., Ahmad, F.: Network intrusion detection system: a systematic study of machine learning and deep learning approaches. Trans. Emerg. Telecommun. Technol. **32**(1), 4150 (2021)
2. Einy, S., Oz, C., Navaei, Y.D.: The anomaly- and signature-based ids for network security using hybrid inference systems. Math. Probl. Eng. **2021**(1), 6639714 (2021). https://doi.org/10.1155/2021/6639714
3. Ahmad, Z., Shahid Khan, A., Wai Shiang, C., Abdullah, J., Ahmad, F.: Network intrusion detection system: a systematic study of machine learning and deep learning approaches. Trans. Emerg. Telecommun. Technol. **32**(1), 4150 (2021). https://doi.org/10.1002/ett.4150
4. Dunmore, A., Jang-Jaccard, J., Sabrina, F., Kwak, J.: A comprehensive survey of generative adversarial networks (GANs) in cybersecurity intrusion detection. IEEE Access **11**, 76071–76094 (2023). https://doi.org/10.1109/ACCESS.2023.3296707
5. Zavrak, S., Iskefiyeli, M.: Anomaly-based intrusion detection from network flow features using variational autoencoder. IEEE Access **8**, 108346–108358 (2020). https://doi.org/10.1109/ACCESS.2020.3001350
6. Kingma, D.P.: Auto-encoding variational bayes. arXiv preprint arXiv:1312.6114 (2013)
7. Goodfellow, I., et al.: Generative adversarial nets. In: Ghahramani, Z., Welling, M., Cortes, C., Lawrence, N., Weinberger, K.Q. (eds.) Advances in Neural Information Processing Systems, vol. 27. Curran Associates, Inc. (2014)
8. Binbusayyis, A., Vaiyapuri, T.: Unsupervised deep learning approach for network intrusion detection combining convolutional autoencoder and one-class SVM. Appl. Intell. **51**(10), 7094–7108 (2021)
9. Hanselmann, M., Strauss, T., Dormann, K., Ulmer, H.: Canet: an unsupervised intrusion detection system for high dimensional can bus data. IEEE Access **8**, 58194–58205 (2020). https://doi.org/10.1109/ACCESS.2020.2982544
10. Ghazdali, A., Hadri, A., Laghrib, A., Nachaoui, M.: Poisson noise and gaussian noise separation through copula theory. Multimedia Tools Appl. **83**(26), 67927–67952 (2024)
11. Wang, W., Jian, S., Tan, Y., Wu, Q., Huang, C.: Robust unsupervised network intrusion detection with self-supervised masked context reconstruction. Comput. Secur. **128**, 103131 (2023)

12. Halvorsen, J., Izurieta, C., Cai, H., Gebremedhin, A.: Applying generative machine learning to intrusion detection: a systematic mapping study and review. ACM Comput. Surv. **56**(10), 1–33 (2024). https://doi.org/10.1145/3659575
13. Kobyzev, I., Prince, S.J., Brubaker, M.A.: Normalizing flows: an introduction and review of current methods. IEEE Trans. Pattern Anal. Mach. Intell. **43**(11), 3964–3979 (2020)
14. Dang, Z., Zheng, Y., Lin, X., Peng, C., Chen, Q., Gao, X.: Semi-supervised learning for anomaly traffic detection via bidirectional normalizing flows. arXiv preprint arXiv:2403.10550 (2024)
15. Ghazdali, A., Ourdou, A., Hakim, M., Laghrib, A., Mamouni, N., Metrane, A.: Robust approach for blind separation of noisy mixtures of independent and dependent sources. Appl. Comput. Harmon. Anal. **60**, 426–445 (2022)
16. Matejek, B., Gehani, A., Bastian, N.D., Clouse, D., Kline, B., Jha, S.: Safeguarding network intrusion detection models from zero-day attacks and concept drift. arXiv preprint arXiv:2403.10550 (2024)
17. Laghrib, A., Nachaoui, A.: A new fractional-order regularization for speckle image denoising: preserving edges and features. Circuits Syst. Signal Process. 1–29 (2025)
18. El Malki, I., El Hakoume, A., Laghrib, A., Hadri, A.: Improved image denoising via self-supervised weickert operator learning and plug-and-play learned primal dual. Neurocomputing **620**, 129267 (2025)
19. Laghrib, A., Bella, F., Nachaoui, M., Hakim, M.: Learning nonlocal weights for second-order nonlocal super-resolution. Discrete Continuous Dyn. Syst.-S **18**(1), 238–268 (2025)
20. Idrissi, M.J., Alami, H., Bouayad, A., Berrada, I.: NF-NIDS: normalizing flows for network intrusion detection systems. In: 2023 10th International Conference on Wireless Networks and Mobile Communications (WINCOM), pp. 1–7. IEEE (2023)
21. Ayed, M.A., Talhi, C.: Federated learning for anomaly-based intrusion detection. In: 2021 International Symposium on Networks, Computers and Communications (ISNCC), pp. 1–8. IEEE (2021)
22. Idrissi, M.J., et al.: Fed-ANIDS: federated learning for anomaly-based network intrusion detection systems. Expert Syst. Appl. **234**, 121000 (2023)
23. Jin, Z., Zhou, J., Li, B., Wu, X., Duan, C.: FL-IIDS: a novel federated learning-based incremental intrusion detection system. Futur. Gener. Comput. Syst. **151**, 57–70 (2024)
24. Rahman, S.A., Tout, H., Talhi, C., Mourad, A.: Internet of things intrusion detection: centralized, on-device, or federated learning? IEEE Network **34**(6), 310–317 (2020)
25. Li, J., Lyu, L., Liu, X., Zhang, X., Lyu, X.: FLEAM: a federated learning empowered architecture to mitigate DDoS in industrial IoT. IEEE Trans. Industr. Inf. **18**(6), 4059–4068 (2021)
26. Engelen, G., Rimmer, V., Joosen, W.: Troubleshooting an intrusion detection dataset: the CICIDS2017 case study. In: 2021 IEEE Security and Privacy Workshops (SPW), pp. 7–12. IEEE (2021)
27. Liu, L., Engelen, G., Lynar, T., Essam, D., Joosen, W.: Error prevalence in NIDS datasets: a case study on CIC-IDS-2017 and CSE-CIC-IDS-2018. In: 2022 IEEE Conference on Communications and Network Security (CNS), pp. 254–262. IEEE (2022)

Classification of Diabetic Retinopathy Anomalies in Fundus Images Using Convolutional Neural Networks

Amine El Hossi[1(✉)], Abdelali Elmoufidi[2], and Mourad Nachaoui[1]

[1] Mathematics Applications Laboratory, Faculty of Sciences and Technics, Sultan Moulay Slimane University, Beni Mellal, Morocco
amine.elhossi@usms.ac.ma, m.nachaoui@usms.ma

[2] Prospective Research Laboratory in Economics and Management, National School of Commerce and Management, Hassan 2 University, Casablanca, Morocco
a.elmoufidi@usms.ma

Abstract. Diabetic retinopathy (DR) is a major cause of vision loss worldwide, emphasizing the need for timely and precise diagnosis. This study introduces a convolutional neural network (CNN) method utilizing EfficientNetB0 as a base network, for classifying DR stages: No DR, Mild DR, Moderate DR, Severe DR and Proliferative DR using fundus images. The model was trained and validated on the APTOS 2019 dataset of retinal images representing various stages of DR, with preprocessing steps applied to enhance critical features. The CNN achieved strong classification performance by utilizing an 80-20 train-test split. Results indicate that the model effectively distinguishes between different stages of DR, providing a valuable tool for automating early detection and aiding ophthalmologists in clinical diagnosis, ultimately reducing diagnostic time and improving patient outcomes. The model achieved 93.5% accuracy, with a specificity of 90.2% and a sensitivity of 91.7%, indicating balanced performance in identifying DR stages while minimizing misclassifications.

Keywords: Diabetic Retinopathy (DR) · Fundus Images · Convolutional Neural Network (CNN) · Deep Learning · Medical Image Classification · Retinal Anomalies

1 Introduction

Diabetic retinopathy (DR) is an illness that occurs as a complication of diabetes mellitus (DM), the scientific name of diabetes. DM is a serious condition characterized by elevated glucose levels in the blood. It arises from the pancreas's inability to produce insulin, or at least a sufficient dose. Or when the body isn't responding effectively to the produced hormone that regulates blood sugar levels. In general, DM is a metabolic disease characterized by having irregular glucose levels in the blood [1]. Diabetes causes serious health complications,

A. Elmoufidi and M. Nachaoui—These authors contributed equally to this work.

A. Laghrib and A. Ghazdali (Eds.): ICMDS 2024, LNNS 1466, pp. 82–91, 2025.
https://doi.org/10.1007/978-3-031-94802-2_8

such as heart disease, blindness, lower limb amputations, and DR [2]. DR is a silent illness that is only detected through its latter stages, sadly when therapy is impossible then. That's why a periodical screen and diagnosis of the eye is very crucial for diabetic patients to have early detection of DR [3]. A healthy eye is visible in the fundus image, where it appears clean and doesn't have any anomalies as shown in the Fig. 1. DR is progressively complicated, from early stages called non-proliferative (NPDR), where diabetic patients generally experience normal vision. In this stage, the retina appears healthy, with no signs of diabetic retinopathy. There are no visible lesions, such as microaneurysms or hemorrhages, indicating that the blood vessels are intact. However, the high glucose level in the blood starts weakening the retinal blood vessels. So they swell and may leak, leading to blurred vision. The more it progresses to proliferative DR (PDR), abnormal, fragile blood vessels are formed, which can cause serious complications like bleeding in the eye, retinal detachment, and even blindness if untreated [4]. DR is classified into two major stages NPDR and PDR. Further, NPDR is classified as mild, moderate, and severe, as shown in the Fig. 2. Figure 3 represents the various stages of DR. From NO DR, as the retina appears healthy, with no signs of diabetic retinopathy. No visible lesions, such as microaneurysms or hemorrhages, indicate that the blood vessels are intact. Next is MILD DR, characterized by microaneurysms, which are small dots in retinal blood vessels. These microaneurysms are the first detectable signs of DR but do not yet cause significant damage or visual impairment. In moderate DR, we can notice more abnormalities in the retina. Along with an increased number of microaneurysms, in addition to retinal hemorrhages, hard exudates, and cotton wool spots. Next stage is Severe DR, where many blood vessels in the retina are blocked, significantly reducing the blood supply. This results in more extensive retinal hemorrhages and microaneurysms. Additionally, the retina may show areas of ischemia (lack of oxygen) due to these blockages, leading to more severe symptoms. The more complicated stage, Proliferative DR, where new abnormal blood vessels begin to grow in response to the ischemia. These new vessels are fragile and prone to leaking, which can cause severe vision problems, such as retinal detachment or vitreous hemorrhage. Without treatment, this stage can lead to blindness [5]. Figure 4 shows the difference between normal seeing and seeing when having DR, highlighting the serious complications of DR. In this paper, we present an approach to classify diabetic retinopathy stages from NODR to PRDR in fundus images using convolutional neural networks. The paper begins with an introduction that highlights the critical issue of diabetic retinopathy and its devastating effects on patients' vision. The second section is the methodology, discussing the dataset used, preprocessing and data augmentation applied techniques, and model development and training. The final section discusses the results.

2 Dataset

The dataset used in this research is the Asia Pacific Tele-Ophthalmology Society 2019 Blindness Detection (APTOS 2019 BD). Publicly available on Kaggle,

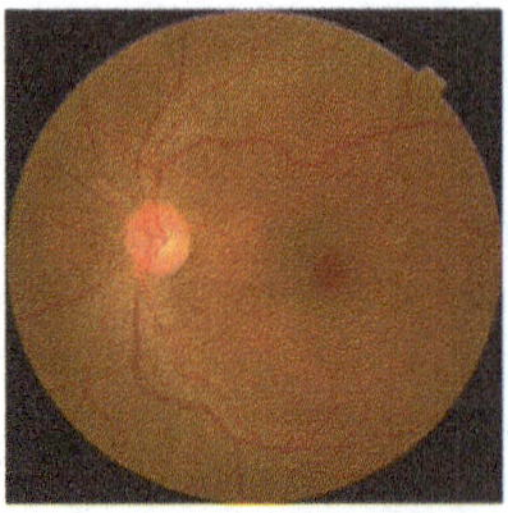

Fig. 1. An example of a normal retinal fundus image. [4]

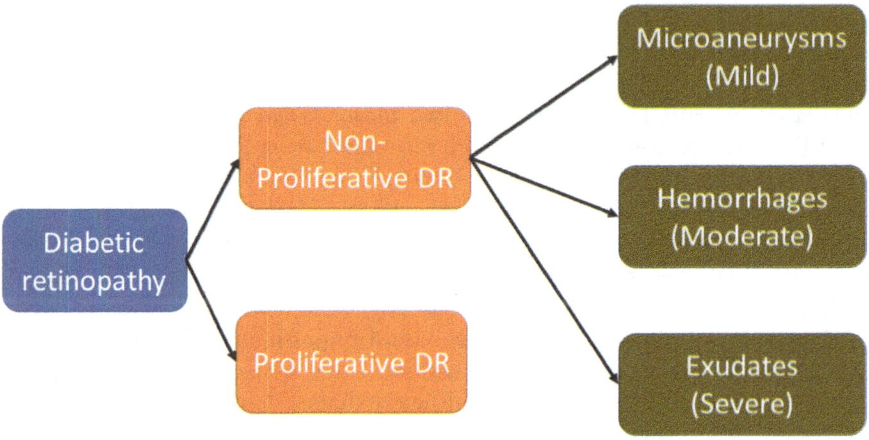

Fig. 2. DR stages [4].

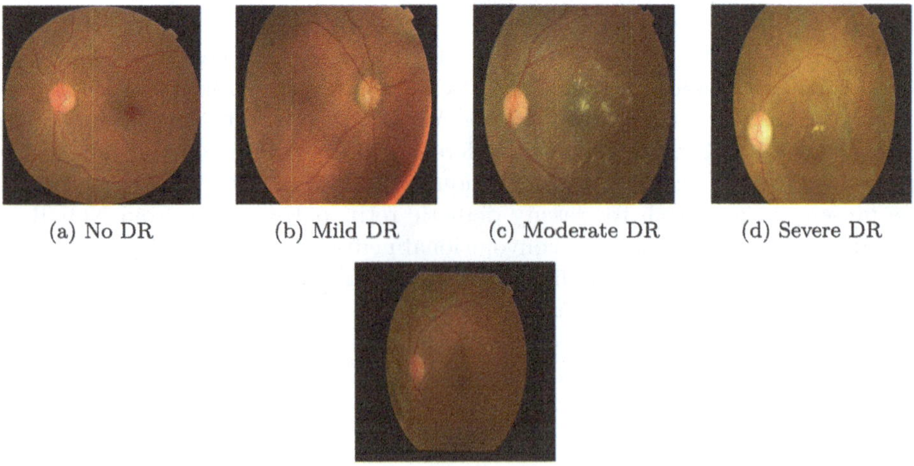

(a) No DR (b) Mild DR (c) Moderate DR (d) Severe DR

(e) Proliferative DR

Fig. 3. Progression of Diabetic Retinopathy from No DR to Proliferative DR [5].

Fig. 4. Normal vision and DR vision.

from the Kaggle Diabetic Retinopathy Detection Challenge 2015, the APTOS dataset contains 3,662 fundus images collected from a broad participant base in rural India. The fundus images were collected over a long period and in varying conditions and environments. Finally, it was labeled by a trained group of doctors into the five categories we saw in the introduction: NO DR (label 0), mild DR (label 1), moderate DR (label 2), severe DR (label 3), and proliferative DR (label 4). As shown in Fig. 5. To ensure robust testing and validation for accurate classification, the dataset is divided into a training set (80%), a testing set (10%), and a validation set (10%) to facilitate model development. The distribution of each set is presented in the Fig. 6.

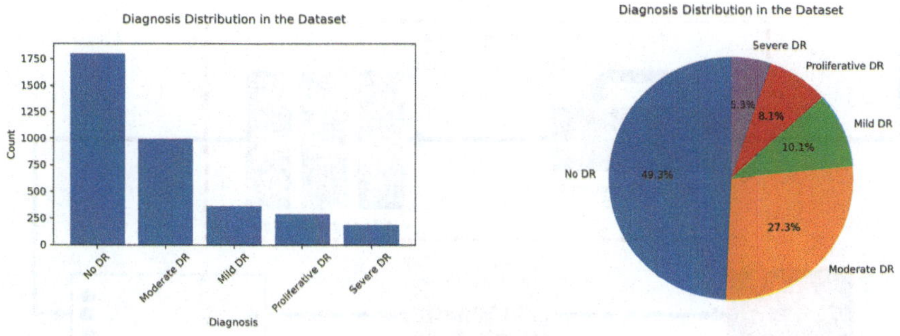

(a) Distribution of Images from each class in the dataset

(b) The percentage of Images in each class relative to the total dataset size

Fig. 5. Diagnosis Distribution in the Dataset.

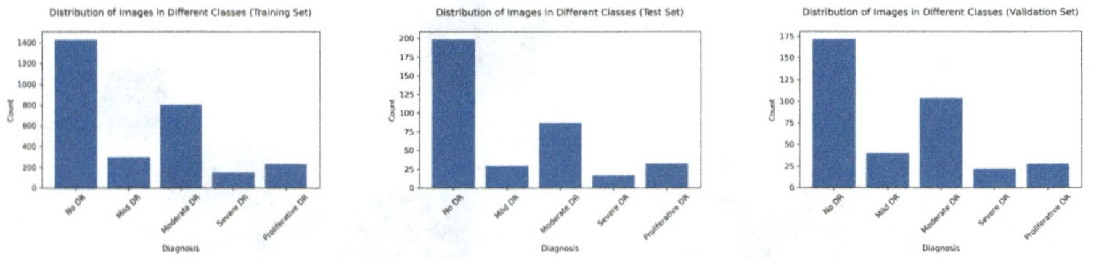

(a) Distribution of Images in the training set. (b) Distribution of Images in the testing set. (c) Distribution of Images in the validation set.

Fig. 6. The distribution of Images in each set.

3 Methodology

In this section, we describe our methodology in detail. As described in the Sect. 2, the APTOS dataset contains 3,662 fundus images with a remarkable variety of quality and brightness across the dataset. Given the quality variation and the limited number of images, our process begins with data pre-processing where a quality enhancement using CLAHE and data augmentation were applied to increase the diversity of the dataset. Creating multiple variations of each image to enhance the robustness of the model. These augmented images are then passed through our proposed model for the training [6–8]. Our methodology aims to make our model able to classify accurately which class the fundus image in input belongs to. The Fig. 7 represents the our methodology flow.

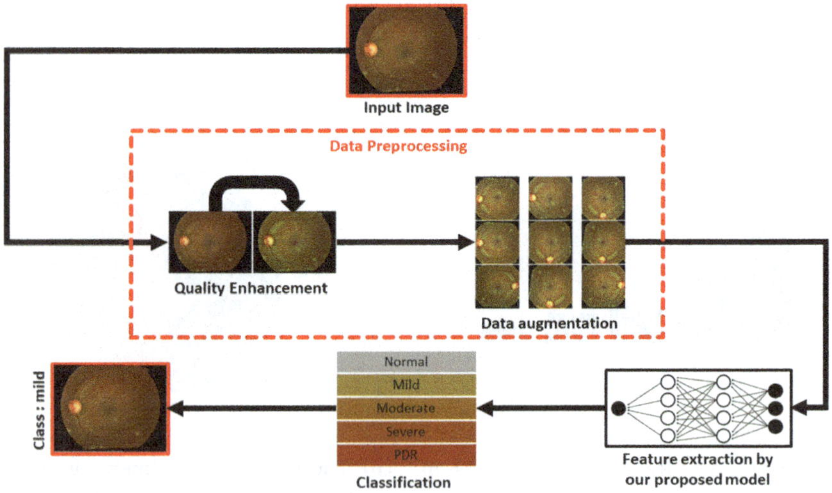

Fig. 7. Proposed methodology diagram flow.

3.1 Data Preprocessing

3.1.1 Quality Enhancement

Our methodology starts with a detailed inspection of the dataset in use. The inspection revealed that the quality of retinal fundus images varies due to differing illumination conditions. Excessive light leads to over-saturation in specific areas and blurs object edges, whereas insufficient light creates dark regions. Fundus images like the Fig. 1, same as any other digital image, consist of three channels (RGB): red, green, and blue. The green channel is found more responsive to applied enhancements than the other channels [9]. To address these low-quality characteristics, an enhancement is applied to the images. Contrast is adjusted using contrast-limited adaptive histogram equalization (CLAHE) on the green channel, the method is presented in Fig. 8. This technique is particularly effective in improving digital images, especially in medical imaging applications, and has shown promising results ([10,11]). An example of a result of the proposed method is shown in Fig. 9. The quality enhancement method makes the features more visible and clear so the kernels of the Convolutional layers perform better in understanding the main difference between each class and fit more accurately on the dataset.

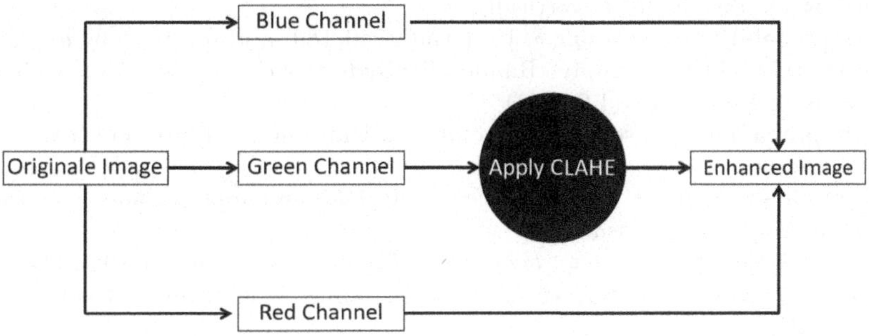

Fig. 8. Applying CLAHE on the green channel for quality enhancement.

3.1.2 Data Augmentation

Deep learning models require a substantial amount of data to avoid overfitting. The fundus images in the APTOS dataset are limited, presenting challenges for training deep learning models. To augment the number of the dataset, we applied various data augmentation techniques. The implementation of data augmentation strategies has been shown to impact the performance of CNN models significantly [12]. These techniques are detailed in Algorithm 1.

3.2 CNN

Artificial neural networks are inspired by biological neurons such as the human brain [13]. They are designed to behave like human cognitive systems. Convo-

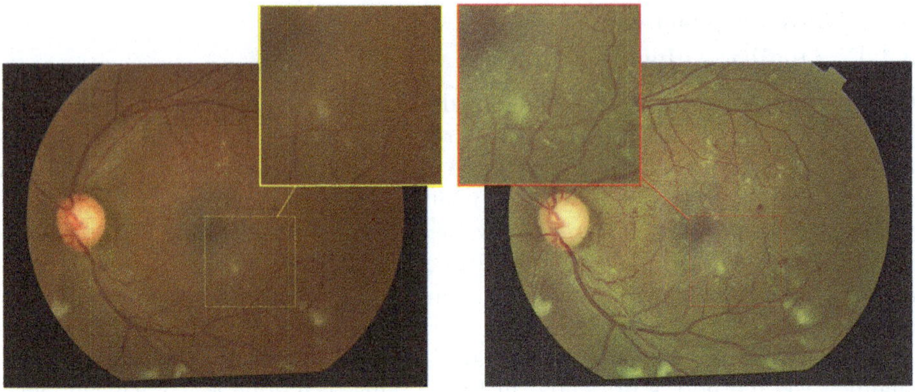

Fig. 9. Comparison of original (left) and enhanced (right) quality image.

Algorithm 1. Proposed data augmentation algorithm

Require: Input fundus image I

Ensure: Augmented image I'

1: $p \Leftarrow 0.75$ ▷ Probability of each transformation
2: $I' \Leftarrow I$
3: With probability p, flip I' vertically
4: With probability p, set $angle \sim$ Uniform$(-180, 180)$ and rotate I' by $angle$
5: With probability 1, apply RandomResizedCrop(I', height=256, width=256, scale=(0.9, 1.0), ratio=(1.0, 1.0))
6: With probability p, set $scale\_factor \sim$ Uniform$(1.0, 1.05)$ and resize I' by $scale\_factor$
7: With probability p, set $scale \sim$ Uniform$(0.0, 0.05)$ and apply perspective distortion to I' with scale factor $scale$
8: With probability p, set $brightness\_factor \sim$ Uniform$(0.01, 0.05)$ and $contrast\_factor \sim$ Uniform$(0.1, 0.1)$, then adjust brightness and contrast of I' accordingly
9: **return** I'

---

lutional neural networks (CNNs) have been proven better than other DL networks in image classification tasks [14]. CNNs are effective especially in computer vision fields when dealing with images because they excel at recognizing patterns within images, such as edges, textures, shapes, and spatial hierarchies. This process makes CNNs particularly adept at image recognition, as they can recognize features regardless of their position within the image. Figure 10 represents an illustration of how CNNs work.

The proposed model is based on the EfficientNetB0 network, pre-trained on the ImageNet dataset, to extract features from input images. EfficientNetB0, being proven much better in classification tasks of medical images [15], is chosen for its optimal balance of model size and accuracy, and modified by removing the

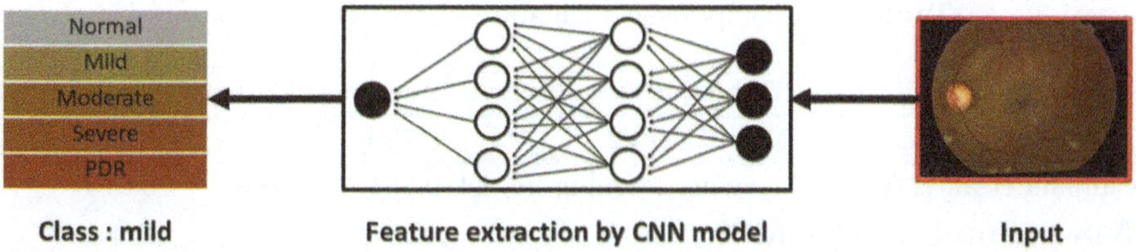

Fig. 10. CNN classification.

top (classification) layer. The Table 1 presents a summary of the model layers and architecture.

Table 1. Model architecture for DR stage classification using the APTOS dataset

Layer	Type	Output Shape	Activation
Base Model	EfficientNetB0 (pre-trained)	(224, 224, 3)	-
Flatten	Flatten	(62720,)	-
Dense_1	Dense	(512,)	ReLU
Dropout_1	Dropout	(512,)	-
Dense_Output	Dense	(5,)	Softmax

4 Results and Discussion

The main objective of our study is the classification of DR stages with CNN model using fundus images. Our model was trained and evaluated on the APTOS dataset, see Sect. 2. The performance of the CNN model was measured through metrics accuracy, sensitivity, and specificity, with results detailed in Table 2. Our approach is compared to existing approaches in Table 3, demonstrating its superior performance and reliability.

Table 2. Metrics tables

Metric	Formula	Value
Accuracy	$\frac{TP+TN}{TP+TN+FP+FN}$	93.5%
Sensitivity	$\frac{TP}{TP+FN}$	90.2%
Specificity	$\frac{TN}{TN+FP}$	91.7%

Table 3. A comparison with state-of-the-art methods.

Study	Technique	Accuracy	Sensitivity	Specificity
Pamadi et al.[16]	MobileNetV2	0.780	-	-
Sanjana et al. [17]	Transfer learning models	0.861	0.854	0.875
Kazakh-British et al.[18]	ANN	0.6	-	-
Our proposed model	EfficientNetB0	0.935	0.902	0.917

5 Conclusion

To conclude, classifying DR anomalies in fundus images using CNNs represents a promising avenue for both research and clinical application. Integrating advanced techniques, effective data augmentation strategies, and exploring various CNN architectures will be crucial in enhancing the accuracy and robustness of automated DR detection systems. In future studies, we will continue to investigate innovative approaches to further improve the classification capabilities of these models, thus enabling early detection and intervention for diabetic retinopathy.

Declarations. The authors declare that the research was conducted in the absence of any commercial or financial relationships that could be construed as a potential conflict of interest.

References

1. Bastaki, S.: Diabetes mellitus and its treatment. Int. J. Diab. Metab. **13**(3), 111–134 (2005)
2. Mukhtar, Y., Galalain, A., Yunusa, U.: A modern overview on diabetes mellitus: a chronic endocrine disorder. Eur. J. Biol. **5**(2), 1–14 (2020)
3. Taylor, R., Batey, D.: Handbook of Retinal Screening in Diabetes: Diagnosis and Management. Wiley (2012)
4. Yasin, S., et al.: Severity grading and early retinopathy lesion detection through hybrid inception-resnet architecture. Sensors **21**(20), 6933 (2021)
5. Porwal, P., et al.: Indian diabetic retinopathy image dataset (IDRID): a database for diabetic retinopathy screening research. Data **3**(3), 25 (2018)
6. Laghrib, A., Bella, F., Nachaoui, M., Hakim, M.: Learning nonlocal weights for second-order nonlocal super-resolution. Discrete Continuous Dyn. Syst.-S **18**(1), 238–268 (2025)
7. Limami, F.-E., Hadri, A., Laghrib, A., Afraites, L.: Fractional optimal control for deep convolutional neural networks exploring ode-based solutions for image denoising. Inverse Probl. Imaging **19**(2), 424–455 (2025)
8. El Malki, I., El Hakoume, A., Laghrib, A., Hadri, A.: Improved image denoising via self-supervised weickert operator learning and plug-and-play learned primal dual. Neurocomputing **620**, 129267 (2025)
9. Setiawan, A.W., Mengko, T.R., Santoso, O.S., Suksmono, A.B.: Color retinal image enhancement using CLAHE. In: International Conference on ICT for Smart Society, pp. 1–3. IEEE (2013)

10. Sahu, S., Singh, A.K., Ghrera, S., Elhoseny, M., et al.: An approach for de-noising and contrast enhancement of retinal fundus image using CLAHE. Opt. Laser Technol. **110**, 87–98 (2019)
11. Li, L., Si, Y., Jia, Z.: Medical image enhancement based on CLAHE and unsharp masking in NSCT domain. J. Med. Imaging Health Inform. **8**(3), 431–438 (2018)
12. Shorten, C., Khoshgoftaar, T.M.: A survey on image data augmentation for deep learning. J. Big Data **6**(1), 1–48 (2019)
13. Abraham, A.: Artificial neural networks. In: Handbook of Measuring System Design (2005)
14. Li, Z., Liu, F., Yang, W., Peng, S., Zhou, J.: A survey of convolutional neural networks: analysis, applications, and prospects. IEEE Trans. Neural Netw. Learn. Syst. **33**(12), 6999–7019 (2021)
15. Padmakala, S.: Deep learning approaches to gastrointestinal image classification: evaluating VGG19 and efficientnetb0 models. In: 2024 7th International Conference on Devices, Circuits and Systems (ICDCS), pp. 205–210 (2024). https://doi.org/10.1109/ICDCS59278.2024.10560955
16. Pamadi, A.M., Ravishankar, A., Nithya, P.A., Jahnavi, G., Kathavate, S.: Diabetic retinopathy detection using mobilenetv2 architecture. In: 2022 International Conference on Smart Technologies and Systems for Next Generation Computing (ICSTSN), pp. 1–5. IEEE (2022)
17. Sanjana, S., Shadin, N.S., Farzana, M.: Automated diabetic retinopathy detection using transfer learning models. In: 2021 5th International Conference on Electrical Engineering and Information Communication Technology (ICEEICT), pp. 1–6. IEEE (2021)
18. Kazakh-British, N.P., Pak, A., Abdullina, D.: Automatic detection of blood vessels and classification in retinal images for diabetic retinopathy diagnosis with application of convolution neural network. In: Proceedings of the 2018 International Conference on Sensors, Signal and Image Processing, pp. 60–63 (2018)

Literature Review of Automatic Text Summarization: Research Trend and Methods

Moaad Elmouktafi(✉), Mourad Jabrane, Amal Ourdou, and Imad Hafidi

Laboratory of Process Engineering, Computer Science and Mathematics, National School of Applied Sciences, Bd Béni Amir, BP 77, 25000 Khouribga, Morocco

moaad.elmouktafi1@usms.ac.ma

Abstract. Text summarization has become an essential tool in managing the rapid expansion of textual information, assisting users in obtaining concise and meaningful summaries of large documents. This literature review explores three primary methodologies in text summarization: extractive, abstractive, and hybrid approaches. Extractive methods focus on selecting key sentences or phrases directly from the original text, while abstractive summarization techniques generate summaries using natural language processing models to rephrase and condense information. Hybrid summarization combines elements of both, aiming to leverage the strengths of each method. This review discusses recent advancements, challenges, and applications within these approaches, emphasizing the growing role of deep learning and language models in transforming text summarization. Finally, recommendations for future research directions are provided, highlighting potential improvements in accuracy, coherence, and scalability.

Keywords: Text Summarization · Abstractive Summarization · Extractive Summarization · Natural Language Processing

1 Introduction

The exponential increase in digital information has profoundly impacted fields like entity resolution [1–4], Sentiment analysis [5] and text summarization [6], intensifying challenges such as scalability, data heterogeneity, and information overload. In this context, an effective summarization of textual content has become essential. From news articles and academic papers to social media and corporate reports, the demand for concise and meaningful representations of information is rapidly growing [7]. Text summarization, an automated process that reduces a text document to its key ideas while preserving the essential message, addresses this need by allowing users to efficiently manage and comprehend large volumes of data.

A. Laghrib and A. Ghazdali (Eds.): ICMDS 2024, LNNS 1466, pp. 92–102, 2025.
https://doi.org/10.1007/978-3-031-94802-2_9

Text summarization can generally be divided into two primary categories: extractive and abstractive. Extractive summarization identifies and selects crucial sentences or phrases directly from the source text, ensuring that the original wording and structure are retained. In contrast, abstractive summarization seeks to understand the content and then generate a summary in new wording, which is closer to human-like summarization. A third, emerging approach, known as hybrid summarization, combines elements of both extractive and abstractive techniques, aiming to improve the quality, coherence, and readability of summaries [8].

Recent advancements in natural language processing (NLP) and machine learning, particularly with the development of large language models (LLMs) [9], have significantly transformed text summarization. These models enable more accurate and contextually aware summarization, particularly for abstractive approaches. This literature review examines the major developments and methodologies within these categories, exploring their strengths, limitations, and practical applications. Additionally, it highlights challenges in the field and offers recommendations for advancing the effectiveness of text summarization technologies.

2 Preliminaries

Text summarization involves reducing a body of text to its essential components while maintaining the main ideas and intent of the original content. This process relies on natural language processing (NLP) techniques to identify and synthesize the key information. In this section, we provide an overview of the fundamental concepts and terminologies central to understanding text summarization, including the primary types of summarization, evaluation metrics, and the technological advancements driving current research in this field.

2.1 Types of Summarization

Text summarization can be broadly categorized into two main types: extractive and abstractive. Extractive summarization involves selecting sentences or phrases directly from the source document based on their relevance, ensuring minimal alteration of the original text [10]. Abstractive summarization, on the other hand, aims to generate new sentences that capture the core ideas, often requiring advanced language modeling and semantic understanding.

2.2 Evaluation Metrics

Evaluating the effectiveness of text summarization is challenging, as it requires balancing informativeness, readability, and conciseness. Common evaluation metrics include ROUGE (Recall-Oriented Understudy for Gisting Evaluation), which measures the overlap of n-grams, word sequences, and word pairs between the generated summary and a reference summary. Other metrics such as BLEU,

METEOR, and BERTScore are also employed to assess summary quality from different perspectives, with increasing focus on contextual coherence and semantic similarity [11].

2.3 Technological Foundations

The evolution of NLP and machine learning has significantly impacted text summarization techniques. Traditional approaches relied heavily on statistical and rule-based methods. However, the advent of deep learning, especially with transformer-based models like BERT, GPT, and T5, has enabled more sophisticated and human-like summarization [9]. These models leverage vast amounts of data to understand context, semantics, and language patterns, making them particularly effective for abstractive and hybrid summarization approaches.

3 Methods

Text summarization, a pivotal task in natural language processing, has evolved significantly, primarily through two approaches: extractive and abstractive methods. Extractive methods select and extract relevant sentences or passages from a source document, often employing statistical, graph-based, or machine learning techniques. Abstractive methods, on the other hand, aim to generate concise summaries by paraphrasing and rewriting source content, typically using neural networks and attention mechanisms. Recent advancements leverage large pre-trained language models, to enhance both extractive and abstractive summarization by capturing context and context-aware summaries. These developments promise improved information retrieval and content understanding in diverse applications.

3.1 Abstractive Summarization Approaches

Abstractive summarization is an advanced approach in natural language processing (NLP) that aims to generate a concise and meaningful summary of a document by interpreting its core ideas rather than simply extracting and reordering sentences [12]. Unlike extractive summarization, which selects and arranges sentences from the source text, abstractive summarization creates new sentences that capture the essence of the text. This technique draws on methods from machine learning, particularly deep learning, and has seen significant improvements with the advent of transformer-based architectures. Abstractive summarization is especially valuable for applications that require brief yet comprehensive content summaries, such as news briefs, legal document summaries, and scientific article abstracts [13].

To illustrate how abstractive summarization works, consider the following example text taken from a dataset on environmental science:

> "Climate change has caused rising temperatures and more frequent extreme weather events worldwide. These changes affect natural ecosystems, human health, and agriculture, putting many species at risk of extinction. Efforts to combat climate change include reducing greenhouse gas emissions, investing in renewable energy, and implementing conservation practices. Public awareness and government policies play a crucial role in addressing these environmental challenges."

Using an abstractive summarization model, this text can be summarized as follows:

> "Climate change leads to global warming and extreme weather, threatening ecosystems and human life. Mitigation strategies include reducing emissions, promoting renewable energy, and implementing conservation efforts, supported by public awareness and policy initiatives."

This summary preserves the core ideas of the original text while rephrasing and condensing the information, showcasing the potential of abstractive summarization to generate coherent, concise, and contextually relevant summaries.

The recent advances in deep learning have led to the development of several models and techniques specifically designed for abstractive summarization. These approaches address challenges like maintaining grammatical accuracy, factual correctness, and fluency in summaries, while also handling issues such as redundancy and out-of-vocabulary (OOV) terms.

3.1.1 Process

Abstractive summarization generally involves four main stages, as illustrated in Fig. 1, each contributing to a more refined and coherent output. These stages are as follows:

1. **Pre-processing:** In the pre-processing phase, the input document is cleaned and tokenized to prepare it for analysis. This step may include removing stop words, punctuation, and irrelevant symbols while standardizing terms to improve model performance. Tokenization divides the document into manageable units, typically words or sub-words, which are then processed by encoding layers in the model. In modern NLP systems, sub-word tokenization is particularly beneficial as it allows the handling of rare or out-of-vocabulary words by breaking them into known sub-components.
2. **Intermediate Representation:** Once pre-processed, the document is converted into an intermediate representation. In this stage, methods such as word embeddings (e.g., Word2Vec or GloVe) or more sophisticated techniques like transformer-based embeddings (e.g., BERT, T5) are used to represent the semantic meaning and context of the words and phrases within the document. These embeddings capture complex relationships and contextual dependencies, allowing the model to understand the underlying themes and structures within the text. This representation serves as the foundation for the next

phase, enabling the model to focus on key information when generating the summary.

3. **Summary Generation:** The summary generation phase is the core of the abstractive summarization process. Using sequence-to-sequence models, often with attention mechanisms, the model identifies and focuses on the most relevant content. Sequence-to-sequence models, such as those based on LSTM (Long Short-Term Memory) networks, are well-suited for this task as they can generate coherent text by processing sequential data. With the introduction of transformers, models like T5, BART, and PEGASUS have further improved this process by providing more sophisticated mechanisms for learning contextual relationships and dependencies. The attention mechanism, in particular, allows the model to assign varying levels of importance to different parts of the text, ensuring that the summary captures critical information while maintaining logical flow.
4. **Post-processing:** In the final stage, post-processing ensures that the generated summary is grammatically correct, fluent, and relevant to the original input. Post-processing techniques may involve sentence restructuring, removal of redundant phrases, and reordering for coherence. This step is crucial for delivering a polished summary that is not only accurate but also easy to read [14]. Additionally, models may incorporate error correction techniques to address common issues like hallucinations (inaccurate or non-existent information) and grammatical inconsistencies that can arise during generation.

3.1.2 Attention-Based LSTMs for Abstractive Summarization

One of the early approaches in abstractive summarization was the attention-based LSTM model, which follows an encoder-decoder architecture. Developed in 2015, this method employs an encoder to process input text and transform each word into a distributed representation through an embedding layer. The model then combines these representations with the hidden states from previous words, forming a sequential understanding of the input text.

In this setup, the decoder generates the output summary word by word. Each word is produced by feeding the previously generated word and the end-of-sequence symbol (EOS) into the decoder. The attention mechanism computes a weight for each word in the input, allowing the model to emphasize relevant sections of the source text. This mechanism, known as global attention, enables

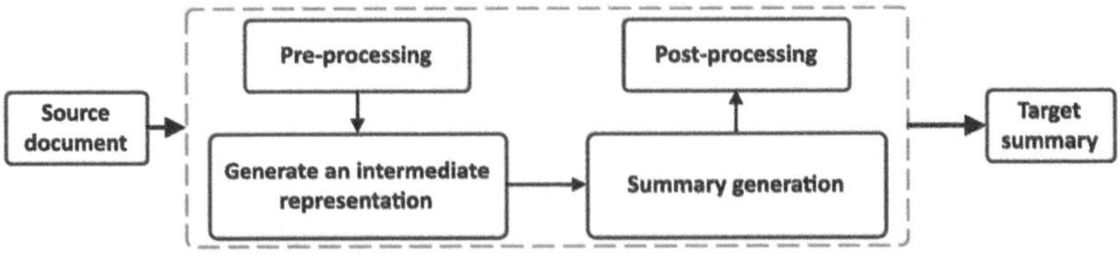

Fig. 1. Abstractive summarization workflow.

the model to consider all input words when generating each word in the output, enhancing coherence and contextual relevance.

The performance of attention-based LSTM models is often evaluated using log loss, a metric that calculates the accuracy of the generated words compared to the original input:

$$-\log p(y_1, \ldots, y_{T'} | x_1, \ldots, x_T) = -\sum_{t=1}^{T'} \log p(y_t | y_1, \ldots, y_{t-1}, x_1, \ldots, x_T)$$

Here, y represents the output words, while x represents the input words.

3.1.3 Pointer-Generator Networks

In 2017, the Pointer-Generator Network was introduced to tackle some limitations of sequence-to-sequence models, such as dealing with OOV words and maintaining factual accuracy. This architecture combines extractive and abstractive techniques, allowing the model to generate new words and directly copy from the source text. This dual functionality is particularly useful for retaining specific terms or facts from the input that may not appear in the models vocabulary.

The Pointer-Generator Network also introduces a coverage mechanism that helps reduce redundancy by tracking previously covered portions of the text. This mechanism informs the attention layers about which sections have already been attended to, reducing the likelihood of repeated content and improving summary coherence.

3.1.4 Advances with Pretrained Transformers

Pretrained transformers represent a significant leap forward in the field of abstractive summarization. Models like BERTSUM, T5, BART, PEGASUS, and BigBird leverage large-scale pretraining on vast text corpora to handle both extractive and abstractive summarization tasks effectively.

BERTSUM is an adaptation of the BERT model tailored for extractive summarization. By adding a classifier that predicts the importance of each sentence, BERTSUM identifies sentences most relevant for inclusion in the summary [15]. Similarly, the T5 model treats every NLP task as a text-to-text problem, making it particularly adaptable to summarization. It was trained on the C4 dataset (Colossal Clean Crawled Corpus) and excels in generating concise and contextually accurate summaries for both extractive and abstractive tasks [16]. BART and PEGASUS offer additional approaches to summarization: BART uses a bidirectional encoder like BERT and an autoregressive decoder similar to GPT, while PEGASUS, specifically designed for summarization, employs a unique pretraining objective where key sentences are masked and generated by the model. For handling longer texts, BigBird extends the transformer architecture with sparse attention patterns, making it effective for summarizing lengthy documents [17]. Large-scale language models like GPT-3 and GPT-4 are also capable of generating human-like summaries with minimal task-specific training and can be fine-tuned on summarization data, enhancing their ability to create coherent and contextually accurate summaries [9].

3.2 Extractive Summarization Approaches

Extractive summarization is a technique in natural language processing (NLP) that generates a summary by directly selecting key sentences or phrases from the original text [18]. Unlike abstractive summarization, which rewrites and synthesizes content, extractive summarization retains exact sentences or words from the input document. This method is widely used for summarizing large bodies of text quickly and efficiently, particularly in domains where precision is critical, such as legal or financial documents.

To demonstrate extractive summarization, consider the following example text taken from a dataset on health and wellness:

> "Regular exercise has been shown to improve physical health, boost mental well-being, and reduce the risk of chronic diseases. Physical activities, such as walking, running, and strength training, contribute to better cardiovascular health, muscle strength, and flexibility. Additionally, exercise helps in managing stress, anxiety, and depression, making it an essential aspect of a healthy lifestyle."

the key sentences extracted could be:

> "Regular exercise improves physical health and mental well-being. Physical activities contribute to cardiovascular health, strength, and flexibility."

In this example, the summary retains sentences that encapsulate the main points, providing an efficient, informative overview without rephrasing or synthesizing new content.

3.2.1 The Extractive Summarization Process

Extractive summarization follows the same workflow illustrated in Fig. 1, and involves several key steps to identify and select the most relevant content from a document:

1. **Pre-processing**: Cleaning and segmenting the input text into manageable units, typically sentences or phrases, for further analysis.
2. **Text Representation**: Creating a structured representation of the document to facilitate text analysis. Common methods include bag-of-words, term frequency-inverse document frequency (TF-IDF), or graph-based representations.
3. **Sentence Scoring**: Assigning a score to each sentence based on its relevance and importance, using selected features.
4. **High-Score Sentence Extraction**: Selecting the top-scoring sentences that collectively summarize the main ideas from the document.
5. **Post-processing**: Reordering sentences, replacing pronouns with antecedents, and refining for coherence and readability.

3.2.2 Diversity-Based Extractive Summarization

Diversity-based selection in extractive summarization aims to avoid redundancy by selecting sentences that represent unique ideas in the text. The Maximal Marginal Relevance (MMR), balances importance and diversity. In MMR, the importance of each sentence is calculated based on a weighted combination of document features. This scoring considers factors such as sentence centrality, similarity to the title, keyword relevance, and proximity to introductory sentences.

$$\text{impr}(s_i) = \text{avg}(\text{WSS}(s_i) + \text{SC}(s_i) + \text{SS\_NG}(s_i) + \text{sim\_fsd}(s_i) + \text{kwrd}(s_i))$$

where **WSS** measures alignment of sentence words with document content, **SC** evaluates sentence centrality within the document, **SS_NG** considers sentence similarity based on n-grams, **sim_fsd** assesses similarity to the documents first sentence and **kwrd** Evaluates the presence of keywords relevant to the document's themes.

3.2.3 Swarm-Based Summarization

Swarm-based methods leverage optimization algorithms like Particle Swarm Optimization (PSO) to enhance feature weighting and improve sentence selection for summarization. In this method, sentences are ranked based on feature scores, with the top-ranked sentences chosen for the summary. PSO uses particles, each representing a potential solution with unique feature weights applied to sentence scoring. The fitness function, often ROUGE-1, evaluates the quality of the generated summary by comparing unigrams with a reference summary.

Recent advancements in semantic-enhanced extractive summarization incorporate distributional semantic models such as Word2Vec to capture word-level semantic relationships. Word embeddings generated by Word2Vec allow models to prioritize semantically relevant sentences, focusing on meaningful content over purely statistical features. For example, in a sentence s_i with features $F = \{\text{WSS, SC, SS\_NG, sim\_fsd, kwrd}\}$, the semantic model adjusts feature scores contextually, ensuring the selection of sentences that better align with the document's core topics.

4 Discussion

In this section, we delve into a comprehensive analysis of extractive and abstractive summarization techniques, exploring their unique strengths, limitations, and associated challenges in the context of natural language processing (NLP). Summarization, an essential task within NLP, aims to distill extensive information into concise and coherent summaries, enabling quick comprehension while preserving the source's key points. Extractive and abstractive summarization approaches offer distinct methodologies to achieve this goal, each with its own advantages and constraints. This section outlines the core characteristics of both approaches, evaluating their effectiveness across different applications and contexts. A comparative analysis table is also provided to encapsulate these findings, highlighting critical aspects for each method.

Extractive summarization identifies and selects relevant phrases or sentences directly from the source text. Unlike abstractive summarization, which generates novel sentences, extractive summarization uses statistical, graph-based, or machine learning techniques to choose the most representative sentences, creating summaries by extracting content verbatim from the original document. Extractive summarization offers several strengths. It is often simpler and more computationally efficient, making it suitable for applications requiring precise, contextually accurate extraction. Additionally, it retains the original meaning by directly using sentences from the source, reducing the risk of inaccuracies, which is particularly beneficial in domains such as legal or technical documents. Furthermore, it demands fewer computational resources compared to abstractive methods, especially when traditional techniques like TF-IDF or graph-based algorithms are employed. However, extractive methods have notable weaknesses. Extracted sentences may lack fluency and logical flow when combined, resulting in disjointed summaries. They are prone to redundancy, often including similar sentences without generating new insights. Moreover, extractive approaches are constrained to the original phrasing and structure, limiting their flexibility in applications requiring brevity or higher-level interpretation. Extractive summarization also faces challenges in achieving contextual understanding, as it focuses on sentence-level importance rather than the document's overall thematic representation. Sentence scoring and selection can also be problematic, especially for documents with complex structures or overlapping content.

On the other hand, abstractive summarization is a more advanced approach in NLP, aiming to interpret the source text and generate summaries with new phrasing that conveys the essence of the document. Abstractive techniques, often relying on deep learning and transformer-based architectures, generate summaries that are concise, coherent, and human-like. This approach offers several strengths, including its ability to provide high-level interpretation, synthesizing information to capture core ideas and improving brevity. Generated summaries tend to be fluent and coherent, making them ideal for applications requiring readability and flow. Additionally, abstractive summarization can handle long and complex documents by distilling information and avoiding redundancy. However, abstractive methods have weaknesses as well. They may introduce errors, such as hallucinated facts or omissions of essential details, due to their generative nature. Abstractive models also require significant computational resources for training and inference, particularly when using large transformer-based models. Furthermore, their performance depends heavily on the quality of training data, with limitations when datasets lack diversity in language structures or domains. Key challenges in abstractive summarization include maintaining factual accuracy to avoid fabricating information, ensuring grammar and fluency for readable output, and managing complexity in long documents to maintain relevance across sections and handle intricate details effectively.

Overall, Table 1 presents a summary of the strengths, weaknesses, and challenges of extractive and abstractive summarization methods.

Table 1. Comparative Analysis of Extractive and Abstractive Summarization Methods

Aspect	Extractive Summarization	Abstractive Summarization
Strengths	- Simple and computationally efficient - Retains original sentence meaning - Effective for technical and legal documents	- Generates high-level interpretations - Coherent and fluent summaries - Effective in generating brief and diverse content
Weaknesses	- May lack coherence between extracted sentences - Can introduce redundancy - Limited flexibility in summarization style	- Potential for factual inaccuracies - Higher computational demand - Performance dependent on training data
Challenges	- Capturing full context - Scoring and selecting sentences appropriately	- Ensuring factual accuracy - Managing grammar and fluency - Handling complex or lengthy documents

5 Conclusion

In conclusion, text summarization continues to play a pivotal role in managing and digesting vast amounts of information across various domains, from news media to academic research and business analytics. This literature review has discussed the primary approaches to text summarization-extractive, abstractive, and hybrid-highlighting the strengths and limitations of each. Extractive methods offer straightforward implementation and high fidelity to the original text but may lack coherence and natural flow. Abstractive methods, while closer to human summarization, are computationally intensive and often require large datasets for training. Hybrid approaches attempt to balance these considerations by combining the advantages of both methods.

Despite significant advancements, challenges remain, particularly in terms of achieving high-quality summaries that are accurate, coherent, and contextually relevant. One key limitation of current summarization techniques, especially those based on deep learning, is their dependency on substantial computational resources and extensive labeled datasets. Moreover, while transformer-based models like BERT and T5 have greatly improved abstractive summarization, issues such as factual consistency and interpretability persist.

References

1. Jabrane, M., Hafidi, I., Rochd, Y.: An Improved Active Machine Learning Query Strategy for Entity Matching Problem, pp. 317–327. Springer (2023). https://doi.org/10.1007/978-3-031-29313-9_28
2. Mourad, J., Hiba, T., Yassir, R., Imad, H.: Erabqs: entity resolution based on active machine learning and balancing query strategy. J. Intell. Inf. Syst. (2024). https://doi.org/10.1007/s10844-024-00853-0

3. Jabrane, M., Tabbaa, H., Hadri, A., Hafidi, I.: Enhancing entity resolution with a hybrid active machine learning framework: strategies for optimal learning in sparse datasets. Inf. Syst. **125**, 102410 (2024). https://doi.org/10.1016/j.is.2024.102410
4. Jabrane, M., Toulaoui, A., Hafidi, I.: Enhancing semantic web entity matching process using transformer neural networks and pre-trained language models. Comput. Inform. **43**(6), 1397–1415 (2024). https://doi.org/10.31577/cai_2024_6_1397
5. Jbel, M., Jabrane, M., Hafidi, I., Metrane, A.: Sentiment analysis dataset in Moroccan dialect: bridging the gap between Arabic and Latin scripted dialect. Lang. Resour. Eval. (2024). https://doi.org/10.1007/s10579-024-09764-6
6. Karnik, M.P., Kodavade, D.V.: Abstractive summarization with efficient transformer based approach. Int. J. Recent Innov. Trends Comput. Commun. **11**(4), 291–298 (2023). https://doi.org/10.17762/ijritcc.v11i4.6454
7. Islam, M.N., Mim, S.T., Tasfia, T., Hossain, M.M.: Enhancing patient treatment through automation: the development of an efficient scribe and prescribe system. Inform. Med. Unlocked **45**, 101456 (2024). https://doi.org/10.1016/j.imu.2024.101456
8. Dewi, K.E., Widiastuti, N.I.: The design of automatic summarization of Indonesian texts using a hybrid approach. Jurnal Teknologi Informasi dan Pendidikan **15**(1), 37–43 (2022). https://doi.org/10.24036/jtip.v15i1.451
9. Lewis, M., Liu, Y.: BART: denoising sequence-to-sequence pre-training for natural language generation, translation, and comprehension. In: Proceedings of the 58th Annual Meeting of the Association for Computational Linguistics, pp. 7871–7880. Association for Computational Linguistics, Online (2020). https://doi.org/10.18653/v1/2020.acl-main.703
10. Nidhi, M., Farhan, K., Amit, M.: Revolutionizing text summarization: a breakthrough in content compression. Int. J. Perform. Eng. **20**(1), 40 (2024). https://doi.org/10.23940/ijpe.24.01.p6.4047
11. Ma, Y., et al.: Text visualization for geological hazard documents via text mining and natural language processing. Earth Sci. Inf. **15**(1), 439–454 (2022). https://doi.org/10.1007/s12145-021-00732-0
12. Berezin, S., Batura, T.: Named entity inclusion in abstractive text summarization (2023). https://doi.org/10.48550/ARXIV.2307.02570
13. Lam, K.N., Doan, T.G., Pham: abstractive text summarization using the brio training paradigm (2023). https://doi.org/10.48550/ARXIV.2305.13696
14. El Malki, I., El Hakoume, A., Laghrib, A., Hadri, A.: Improved image denoising via self-supervised weickert operator learning and plug-and-play learned primal dual. Neurocomputing **620**, 129267 (2025)
15. Liu, Y., Lapata, M.: Text summarization with pretrained encoders. In: Inui, K., Jiang, J., Ng, V., Wan, X. (eds.) Proceedings of the 2019 Conference (EMNLP-IJCNLP), pp. 3730–3740. Association for Computational Linguistics, Hong Kong, China (2019). https://doi.org/10.18653/v1/D19-1387
16. Gupta, S.: A hybrid approach to single document extractive summarization (2018). https://api.semanticscholar.org/CorpusID:212483094
17. Zaheer, M., Guruganesh, G., Dubey, K.A.: Big bird: transformers for longer sequences. In: Proceedings of the 34th International Conference on Neural Information Processing Systems. NIPS 2020. Curran Associates Inc., NY, USA (2020)
18. Mittal, N., Agarwal, B., Mantri, H., Goyal, R.K., Jain, M.K.: Extractive text summarization (2014)

Prior-Guided Image Denoising via Generative Adversarial Networks with Single Noisy Images

Hind Hafsi(✉) and Abdelghani Ghazdali

LIPIM, ENSA Khouribga, USMS, Khouribga, Morocco
hind.hafsi@usms.ac.ma, a.ghazdali@usms.ma

Abstract. In order to address problems with image denoising, numerous approaches have been developed. Generative models arose as one of the most extensively researched areas in machine learning, particularly deep learning. Generative Adversarial Networks, a recent major discovery in the generative subject matter, are very fascinating. This research presents a novel approach for image denoising, notably for single noisy images, based on generative adversarial networks (GAN). The observations and statistics on the corruption process serve as the only source of information. The procedure optimizes the GAN as the prior and seeks to determine the maximum a posteriori (MAP) estimate. For creating realistic reconstructions from a corrupted observation, the generator adds a penalty term to impose the reconstructions on the associated observation. We demonstrate our methodology using CelebA dataset with various sizes and degrees of degradation. The suggested approach provides an alternative to unsupervised denoising and delivers outcomes that are on line with the state-of-the-art in noise reduction for generative models.

Keywords: generative adversarial networks · image denoising · inverse problems · unsupervised learning

1 Introduction

Image denoising continues to be a vital and vibrant research topic in low-level vision, necessitated by its broad range of practical applications [12–15]. The primary challenge involves reconstructing a clean image x from its noisy counterpart y, typically expressed by the degradation model $y = x + v$, where v signifies additive noise. Understanding the nature of this noise and effectively modeling image priors is crucial, especially from a Bayesian perspective [29].

Historically, a variety of statistical methods have been employed to tackle image noise, primarily utilizing modified Gaussian-like filters to smooth pixel values. Although these methods help in reducing noise, they often introduce unwanted blurring. With the advent of deep learning, a shift towards data-driven approaches has been observed. These modern algorithms leverage large datasets

A. Laghrib and A. Ghazdali (Eds.): ICMDS 2024, LNNS 1466, pp. 103–112, 2025.
https://doi.org/10.1007/978-3-031-94802-2_10

of noisy-clean image pairs, enabling the reconstruction of high-quality images from noisy observations [6,7,16].

One significant limitation of deep learning methods is their dependency on extensive, high-quality datasets, which are both expensive and challenging to compile [11]. Recently, unsupervised learning has offered new perspectives, particularly through the use of GANs, which allow for the generation of clean images from noisy data without direct clean examples [22,32].

In this paper, we explore how advancements in neural network technologies are addressing the persistent challenge of image denoising. We highlight our innovative approach using a deep generative model that operates effectively with only a single noise sample, circumventing the traditional requirement for paired data sets. This method not only simplifies the learning process but also demonstrates promising results in producing clean images from noisy inputs.

2 Background

Image denoising continues to be an important and evolving research topic in the fields of machine learning and image processing, with each method bringing its own challenges, assumptions, advantages, and disadvantages. This section offers a review of major works on image denoising.

Traditional approaches to image denoising have included the use of pixel-wise filters, which are broadly classified into non-linear and linear types. Non-linear filters, which operate without explicitly identifying the noise [19,20], typically apply low-pass filtering techniques to address high-frequency noise components. Although effective at reducing noise, these filters are known to cause image blurring [5,17,30].

Linear filters, on the other hand, are exemplified by the mean filter, which is particularly effective against Gaussian noise in terms of minimizing mean square error. However, these filters often fail to preserve critical image features such as lines and sharp edges and are generally ineffective against signal-dependent noise. The Wiener filter, which is optimal for smooth signals, necessitates prior knowledge of the noise characteristics and the original signal's spectrum [4,18]. To address these limitations, wavelet denoising has been introduced, exploiting the sparsity and multiresolution structure inherent in wavelet transforms. This has given rise to sophisticated techniques ranging from multiresolution wavelet coefficient thresholding [26] to statistical wavelet analysis [24].

The advent of deep learning has significantly impacted various fields, including image denoising, by employing multi-layer architectures to extract higher-level features from raw data [25]. Post-2015, the deployment of deep autoencoders (AE) and variational autoencoders (VAE) marked a notable advancement in denoising techniques, although they occasionally produce outputs with a slight blur [1,23].

GANs, introduced by Ian Goodfellow in 2014 [22], have revolutionized the approach to modeling and sampling image distributions. The shift towards GAN-based denoising solutions has yielded cutting-edge results, initially utilizing GANs to simulate noise distributions to augment datasets, and subsequently employing them in supervised learning frameworks to denoise images using pairs of noisy and clean images [28].

Further investigations have utilized generative models to resolve ill-posed inverse problems by first training on a distribution of undistorted signals. The task then involves reconstructing the signal from a corrupted measurement, necessitating extensive computational effort, including solving complex optimization problems that often require significant GPU resources and multiple restarts to avoid suboptimal local minima [2,31].

3 Deep Generative Models

Developed by Ian Goodfellow in 2014, GANs are generative models that generate samples from a distribution, unlike discriminative models, which identify if a sample belongs to a distribution. Training a GAN marked a major advancement, though GANs evolved from prior foundational models.

AE are also generative models and precede GANs. AEs, attributed to Rumelhart et al. [9,27] in the 1980s, utilize neural networks with a bottleneck to represent inputs in a compressed form and then reconstruct them, enabling generation of new samples. In training, AEs compress images into a lower-dimensional space, then attempt to recreate the original image, often using mean squared error as a loss measure. During testing, the generator portion of the AE creates new samples, though basic AEs have limitations like blurriness and poor detail representation. Variants like VAEs aim to reduce these issues, but some lower-dimensional space outputs remain nonsensical [10].

A GAN architecture consists of two networks: a generator and a discriminator. The generator creates samples from a seed input, while the discriminator differentiates between real and generated data. The discriminator loss uses binary cross entropy, and the generator optimizes to make its outputs appear real to the discriminator. This adversarial process refines the generator, leading it to produce samples that increasingly resemble the true data distribution.

$$\min_G \max_D V(D,G) = \mathbb{E}x \sim pdata(x)[\log(D(x))] + \mathbb{E}_{z \sim p_z(z)}[1 - \log(D(G(x)))] \quad (1)$$

In ideal conditions, a Nash equilibrium between the generator and discriminator may be reached, where the generated distribution closely matches the actual data distribution [22].

4 Our Method

Let $X \sim P_X$ be the clean image that we want to recover, Let $Y \sim P_Y$ be noisy measurement. We also suppose that we can only get to the clean image via the noisy distribution. Let F be a the degradation/measurement function that gets items from X and outputs their measurements in y. Next suppose that we only have access to a subset of the noisy measurement $\hat{Y} \subset Y$.

For any $\in Y$ we would like to reconstruct $\hat{x}$ s.t. $y = F(\hat{x})$. In a broader sense, $Y = \{F(x)|x \in X\}$.

Representing the inverse problem as finding the MAP estimate, we can described as fellow:

$$\hat{x} = \arg\max_{x} \; p_{Y|X}(y|x) \; . \; p_X(x) \tag{2}$$

$p_{Y|X}(y|x)$ denotes the likelihood term and p_X (x) denotes the prior. The likelihood indicates the probability that y is the outcome of a corruption measurement on x, whereas the prior indicates the certainty that x belongs to the real image distribution.

To acknowledge the true image, our algorithm will use a GAN model. We invent the issue of developing a generator $G : Y \rightarrow X$ such that the generated output is the related MAP estimate $\hat{x}$ for each input y. However, unlike earlier work, we adopt an analytic form of the likelihood, as will be demonstrated in the next sections.

Therefore we could express the objective as:

$$G_{MAP} = \arg\max_{G} \mathbb{E}_{p_Y} \{\log p_{Y|X}(y|G(y)) + \log p_X(G(y))\} \tag{3}$$

Equation (2) was converted to Eq. (3) by substituting G(y) for x and taking the equation's logarithm.

In our method, we suppose that the noise is independent and identically distributed (i.i.d) pixel wise and chosen from a distribution that can be defined analytically in order to represent its likelihood. The likelihood $p_{Y|X}(y|x)$ is then able to be stated in its analytical form, allowing us to maximize over it.

$$p_{Y|X}(y|x) = p(y_0|x_0) \cdot \ldots \cdot p(y_n|x_n) \tag{4}$$

The analytic representation of the likelihood from which we can optimize. It is worth noting that Eq. (4) would be forcing G to provide outputs that are similar to the input of y, rather than just some random instance from the distribution of X.

For example if we have additive zero mean Gaussian noise with standard deviation σ

$$G_{MAP} = \arg\min_{G} \mathbb{E}_{p_Y} \left[-\log(p_X(G(y))) + \lambda \|y - G(y)\|^2\right] \tag{5}$$

Using the cross entropy notion $H(p,q)$, we are able to state the following:

$$G_{MAP} = \arg\min_{G} -\mathbb{E}_{G_Y} \left[\log(p_X(G(y)))\right] + \lambda \mathbb{E}_{p_Y} \left[\|y - G(y)\|^2\right] \tag{6}$$

$$= \arg\min_{G} H(p_{G(y)}, p_X) + \lambda \mathbb{E}_{p_Y} \left[\|y - G(y)\|^2\right] \tag{7}$$

Even though we have comprehensive knowledge of G, handling the cross-entropy $H(p_{G(y)}, p_X)$ based just on training samples is still challenging. Instead, we estimate G_{MAP} using:

$$\hat{G} = \arg\min_{G} KL(p_{G(y)}, p_X) + \lambda \mathbb{E}_{p_Y}\left[\|y - G(y)\|^2\right] \quad (8)$$

Putting the Prior, in contrast, is difficult to express properly, therefore we must rely on machine learning to find a solution. As demonstrated in the literature, an unconditional GAN can be used to derive a prior distribution P_X from a random input distribution P_R.

GAN should be able to find $G(r) = x$ with $r \in R$, in which the probability of x being drawn from P_X is high. In the approach we propose, we choose to denoise certain images rather than randomly selecting an example from the target distribution X. The probability ensures that the output of $G : Y \rightarrow X$ is similar to its input, therefore for the prior component of Eq. (3), we want to optimize that the output of G is an item from the distribution of X.

So next, it's discussing the GANs discriminator D loss function and how it connects to the generator's loss function. The expressed objective of the discriminator is:

$$\hat{D} = \arg\max_{D} \mathbb{E}_{p_Y}\{\log D(y) + \log(1 - D(F(G(y))))\} \quad (9)$$

where $\hat{D}$ is the best discriminator. This enables us to get an ideal generator within our solution architecture by optimizing the discriminator. For our purposes, an ideal generator is the one that can reconstruct the original clean distribution with accuracy.

We demonstrated that an expression of likelihood can be identified and optimized. Similarly, we demonstrated that even with only noisy data, it is possible to get generator to output examples from a clean distribution.

In order to enforce the likelihood term, we must look into the noise distribution's unique analytic expression. For Gaussian noise, we can see that the loss is written as

$$loss_{likelihood}(G) = \mathbb{E}_{p_Y}\left[\|y - G(y)\|^2\right] \quad (10)$$

Recalling from Eq. (9) that D is the best discriminator, we can conclude that for the prior generator loss:

$$loss_{prior}(G) = \mathbb{E}_{p_Y}\{\|y_G(y)\|^2\} \quad (11)$$

Considering both penalties Eqs. (10) and (11) combined, our total objective function is eventually given as

$$loss_{tota}(G) = \lambda . loss_{likelihood}(G) + loss_{prior}(G) \quad (12)$$

The algorithm of our proposed denoising GAN is described below.

Algorithm 1. Training Process algorithm

Require: Initialize parameters of the generator G and the discriminator D.
  for number of training iterations **do**
    for number of batches **do**
      for k steps **do**

$Y_{real} \leftarrow sample\_batch(Y)$
$Y_{fake} \leftarrow F(G(sample\_batch(Y)))$
Update D by ascending its stochastic gradient:

$$\nabla_{\theta_d} \frac{1}{n} \sum_{i=1}^{n} \{\log D(y) + \log(1 - D(F(G(y))))\} \tag{13}$$

      end for
      $Y_{fake} \leftarrow sample - batch(Y)$
      Update G by descending its stochastic gradient: g

$$\nabla_{\theta_g} \frac{1}{n} \sum_{i=1}^{n} \lambda \{\|y - G(y)\|^2\} + \{1 - \hat{D}(F(G(y)))\} \tag{14}$$

    end for
  end for

5 Experiments

This section contains the results of our experiments. To validate our technique, we mostly employ CelebA image datasets, CelebA: Celebrity data set with around 200 000 samples [33]. [21]'s GAN design serves as a template for our network architectures. We propose an image-to-image version of their generator utilizing for the reconstruction network G. To put ourselves in the closest to actual situation possible, every image has only ever been corrupted once, there are never many instances of the same image corrupted with different corruption measurement.

For the different degradation processes F used in the experiments:

- Remove_Pixel: This degradation process randomly samples a fraction p of pixels uniformly and sets the related channel values to 0. All the corresponding channel values are set to 0.
- Remove_Pixel Channel: Rather than setting to 0 a pixel for all channels as in Remove-Pixel, one samples a pixel coordinate and a channel, and sets the corresponding value to 0.
- Convolve_Noise: Here $F(x; \theta) := k * x + \theta$, where $*$is the convolution operator and k is a mean filter of size l. For every pixel, noise θ sampled from a zero-mean Gaussian of variance plus to the previous result.

We additionally compared our approach against particular baselines with the objective for removing corruptions.

- Conditional Ambient GAN: This baseline is unsupervised. The setting is the same as for our model, the measurement process F is taken to be known, and only the corrupted counterpart P_Y of the uncorrupted signal distribution P_X is accessible. This baseline combines the aforementioned AmbientGan [3] and CS-GAN [2], two current signal recovery approaches. To generate samples from P_X, an unconditional generator p_X^G is trained using the AmbientGan architecture for each type of measurement method F.
  Given a specific measurement y, the reconstruction $\hat{x}$ is the signal from G that is closest to y. To recover $\hat{x} = G(\hat{z}$, we looking to find the latent code $\hat{z}$, such that $\hat{z} = \arg\min_z \|y - G(z)\|_2^2 \ . \ R(z)$, with R(z) is a regularizing term that enforces the latent code to stay in G's input domain. This objective is optimized using stochastic gradient descent.
- Deep Image Prior [31]: This approach aims to discover a reconstruction from G_ϕ that is close to the measurement given a generator G with randomly initialized weights ϕ and a measurement y. However, this supposes access to the true value θ that corrupted datum $y \ : \ y = F(x;\theta)$. Formally, we optimize the following objective: $\arg\min_\phi \|F(y - G_\phi(z);\theta\|_2^2$. F serves as a mask in this situation, removing the terms related to the pixels in y that have been set to 0.

In what follows, we provide an overview of our results. We first evaluate our model using the CelebA dataset. Then, we provide statistical results utilizing samples from the model outputs and these baselines. In these outcomes. We compare our model to the baselines mentioned before. We give the mean square errors (MSE) between the reconstructed image and the true image x that was used to build the input y, shown in Table 1.

Table 1. Average mean square error (MSE) of our proposed method and the other baselines on Celeb datasets

Method	Proposed	DIP	Cond. A. G.
Remove pixel	0.0489	0.0456	0.0398
Remove pixel channel	0.0439	0.0489	0.0382
Convolve noise	0.0235	0.0198	0.0102

Number-wise, our model performs well. All of the methods are fairly similar when it comes to MSE. Our unsupervised model functions effectively. Furthermore, we notice that results are comparable to and occasionally even superior to those of the other two baselines. This demonstrates that even with simply our likelihood term, the reconstruction might be dependent on the input image.

Therefore, quantitative data provide us with sufficient information about the baselines performances. These results shows the quality of our reconstruction. The Figs. 1, 2, 3, demonstrate reconstructions obtained from three models on the same dataset. On the surface, the reconstructions of our model are compatible with the statistical data and visually superior to the other models. Our model can generate images with good visual quality while being consistent with the underlying uncorrupted data.

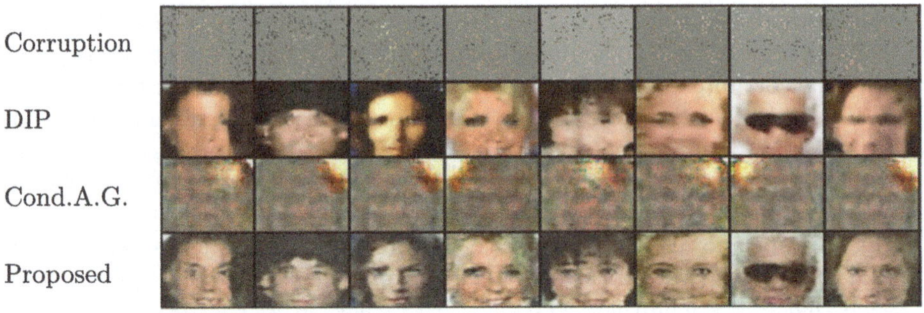

Fig. 1. Baseline comparison for CelebA dataset with Corruption is Remove-Pixel (p = 0.95).

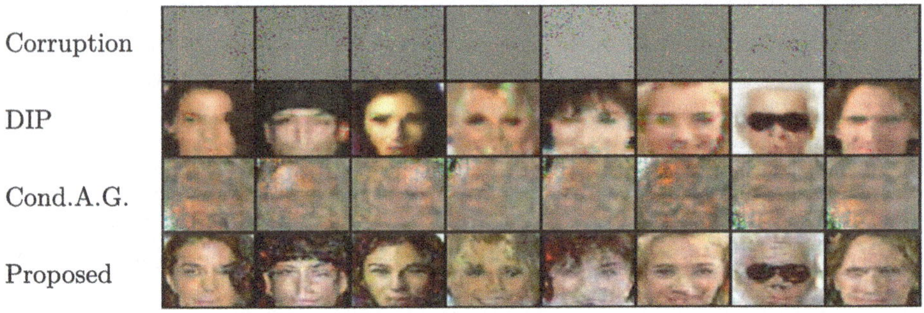

Fig. 2. Baseline comparison for CelebA dataset with Corruption is Remove-Pixel-Channel (p = 0.95).

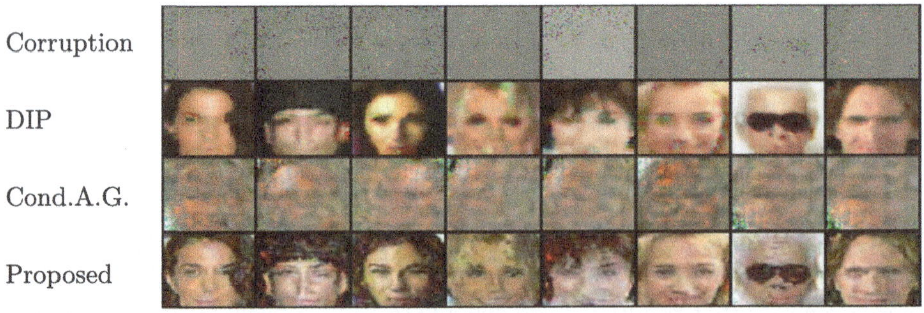

Fig. 3. Baseline comparison for CelebA dataset with Corruption is Convnoise($\sigma_c = 0.2, l = 3$).

6 Conclusion

In this study, we proposed a framework for reconstructing clean images from degraded measurements using a neural network, even without access to uncorrupted image data. We approached the problem by defining each training sam-

ple's task as estimating the most probable clean image based on its observed counterpart. To achieve this, we developed an objective function that combines two key elements: an adversarial loss to enhance image realism and a reconstruction loss to anchor each result to its original observation. Results from our method are comparable to previous baselines from state-of-the-art of unsupervised deep learning models. There are several promising directions for further research and detailed analysis arising directly from this work. One possibility is to continue refining the deep network architecture of our model and experimenting with different training schedulers and learning rate strategies. Future studies could also explore applying our framework to other types of noise and corruption, assessing its effectiveness across diverse datasets and real-world applications.

References

1. Jabbar, A., Li, X., Omar, B.: A survey on generative adversarial networks: variants, applications, and training. ACM Comput. Surv. **54**(8), 1–49 (2020)
2. Bora, A., Jalal, A., Price, E., Dimakis, A.G.: Compressed sensing using generative models. In: Proceedings of the 34th International Conference on Machine Learning, vol. 70, pp. 537–546 (2020)
3. Bora, A., Price, E., Dimakis, A.G.: AmbientGAN: generative models from lossy measurements. In: International Conference on Learning Representations (2018)
4. Jain, A.K.: Fundamentals of Digital Image Processing, p. 569. Prentice-Hall (1989)
5. Hamza, A., Luque-Escamilla, P., Aroza, M., Roldan, R.: Removing noise and preserving details with relaxed median filters. Math. Imag. Vis. **11**, 161–177 (1999)
6. Sinha, A., Lee, J., Li, S., Barbastathis, G.: Lensless computational imaging through deep learning. Optica **4**(9), 1117–1125 (2017)
7. Zuo, C., et al.: Deep learning in optical metrology: a review. Light Sci. Appl. **11**, 1–54 (2022)
8. Donoho, D.L., Johnstone, I.M.: Ideal spatial adaptation by wavelet shrinkage. Biometrika **81**(3), 425–455 (1994)
9. Rumelhart, D.E., Hinton, G.E., Williams, R.J.: Learning internal representations by error propagation, Technical report, California University San Diego La Jolla Inst for Cognitive (1985)
10. Kingma, D.P., Welling, M.: Auto-encoding variational bayes. In: International Conference on Learning Representations (2014)
11. Wang, G., Ye, J.C., De Man, B.: Deep learning for tomographic image reconstruction. Nat. Mach. Intell. **2**(12), 737–748 (2020)
12. Nachaoui, A., Laghrib, A., Hakim, A.: Mathematical Control and Numerical Applications. Springer (2021)
13. Nachaoui, M., Afraites, L., Hadri, A., Laghrib, A.: A non-convex non-smooth bi-level parameter learning for impulse and gaussian noise mixture removing. Commun. Pure Appl. Anal. **21**(4), 1249–1291 (2022)
14. El Malki, I., El Hakoume, A., Laghrib, A., Hadri, A.: Improved image denoising via self-supervised Weickert operator learning and plug-and-play learned primal dual. Neurocomputing **620**, 129267 (2025)
15. Ghazdali, A., Ourdou, A., Hakim, M., Laghrib, A., Mamouni, N., Metrane, A.: Robust approach for blind separation of noisy mixtures of independent and dependent sources. Appl. Comput. Harmon. Anal. **60**, 426–445 (2022)

16. Laghrib, A., Ghazdali, A., Hakim, A., Raghay, S.: A multi-frame super-resolution using diffusion registration and a nonlocal variational image restoration. Comput. Math. Appl. **72**(9), 2535–2548 (2016)
17. Hakim, M., Ghazdali, A., Laghrib, A.: A multi-frame super-resolution based on new variational data fidelity term. Appl. Math. Model. **87**, 446–467 (2020)
18. Ghazdali, A., Hakim, A., Laghrib, A., Mamouni, N., Raghay, S.: A new method for the extraction of fetal ECG from the dependent abdominal signals using blind source separation and adaptive noise cancellation techniques. Theor. Biol. Med. Model. **12**, 1–20 (2015)
19. Ghazdali, A., Hadri, A., Laghrib, A., Nachaoui, M.: A blind Poisson-Gaussian noise separation using learning copula densities. Cir. Syst. Sig. Process. **42**(11), 6564–6590 (2023)
20. Ghazdali, A., Hadri, A., Laghrib, A., Nachaoui, M.: Poisson noise and Gaussian noise separation through copula theory. Multimedia Tools Appl. **83**(26), 67927–67952 (2024)
21. Zhang, H., Goodfellow, I., Odena D., Metaxas, N.: Self-attention generative adversarial networks (2018)
22. Goodfellow, I.J.: Generative adversarial networks. In: Proceedings Advances in Neural Information Processing Systems, pp. 1406–2661 (2014)
23. Liang, J., Liu, R.: Stacked denoising autoencoder and dropout together to prevent overfitting in deep neural network. In: International Congress on Image and Signal Processing (CISP), pp. 697–701 (2015)
24. Portilla, J., Strela, V., Wainwright, M., Simoncelli, E.: Image denoising using scale mixtures of gaussians in the wavelet domain. IEEE Trans. Image Process. **12**(11), 1338–1351 (2003)
25. Dengand, L., Yu, D.: Deep learning: methods and applications. Found. Trends Sig. Process. **7**(4), 197–387 (2014)
26. Zhang, M., Gunturk, B.K.: Multiresolution bilateral filtering for image denoising. IEEE Trans. Image Process. **17**(12), 2324–2333 (1995)
27. Baldi, P.: Autoencoders, unsupervised learning, and deep architectures. In: Proceedings of ICML Workshop on Unsupervised and Transfer Learning, pp. 37–49 (2012)
28. Yang, Q., et al.: Low-dose CT image denoising using a generative adversarial network with Wasserstein distance and perceptual loss. Opt. Express **37**(6), 1348–1357 (2018)
29. Bernstein, R.: Adaptive nonlinear filters for simultaneous removal of different kinds of noise in images. IEEE Trans. Cir. Syst. **34**(11), 1275–1291 (1987)
30. Yang, R., Yin, L., Gabbouj, M., Astola, J., Neuvo, Y.: Optimal weighted median filtering under structural constraints. IEEE Trans. Sig. Process. **43**(3), 591–604 (1995)
31. Lempitsky, V., Vedaldi, A., Ulyanov, D.: Deep learning: methods and applications. In: IEEE/CVF Conference on Computer Vision and Pattern Recognition, pp. 9446–9454 (2018)
32. Bengio, Y., Yao, L., Alain, G., Vincent, P.: Generalized denoising auto-encoders as generative models. Adv. Neural Inf. Process. Syst. **26** (2013)
33. Liu, Z., Luo, P., Wang, X., Tang, X.: Deep learning face attributes in the wild. arXiv preprint arXiv:1411.7766 (2014)

Bridging Energy Networks: Blockchain Interoperability for Efficient Multi-energy Management

Bentachfine Ilyas(✉) and Erritali Mohammed

Data4Earth Laboratory, Sultan Moulay Slimane University, Av Med V, BP 591, 23000 Beni Mellal, Morocco
{ilyas.bentachfine,m.erritali}@usms.ma

Abstract. Blockchain technology has revolutionized energy management in multi-energy systems by enabling secure, transparent, and decentralized transactions. However, its adoption is hindered by significant interoperability challenges, such as isolated platforms and cross-chain communication barriers, which limit scalability and efficiency. This chapter explores the role of blockchain interoperability in addressing these challenges, with a focus on frameworks like BAILIF and cross-chain smart contracts. These solutions enable seamless integration of diverse energy sources, real-time energy trading, and enhanced grid resilience. Case studies of platforms like Power Ledger and Energy Web Chain illustrate practical applications, highlighting the potential for blockchain interoperability to drive innovation in decentralized energy markets.

Keywords: Blockchain Technology · Interoperability · Decentralized Energy Systems · Peer-to-Peer (P2P) Energy Trading · Renewable Energy Integration

1 Introduction

As the global energy sector transitions toward sustainability, the increasing adoption of Distributed Energy Resources (DERs), such as solar, wind, and battery storage systems, presents unprecedented challenges to traditional, centralized energy systems. These systems, originally designed for unidirectional power flows, often struggle to integrate diverse renewable sources, leading to inefficiencies in energy distribution and resource utilization [1,2]. Addressing the complexities of balancing supply and demand across multi-energy systems is critical to realizing the full potential of renewable energy.

Blockchain technology, characterized by its decentralized, transparent, and secure transaction mechanisms, has emerged as a transformative tool for modernizing energy markets [3]. By enabling peer-to-peer (P2P) energy trading and fostering trustless operations, blockchain provides a pathway to overcome many

E. Mohammed—These authors contributed equally to this work.

A. Laghrib and A. Ghazdali (Eds.): ICMDS 2024, LNNS 1466, pp. 113–123, 2025.
https://doi.org/10.1007/978-3-031-94802-2_11

limitations of traditional energy systems. However, its adoption in multi-energy systems is hindered by significant interoperability challenges. Isolated blockchain platforms, each operating independently with unique protocols and standards, create "data silos" that impede seamless collaboration and cross-chain energy trading [4,5].

Addressing blockchain interoperability is crucial to unlocking its full potential in energy systems. Recent advancements, such as the Blockchain Agnostic Interoperability Framework (BAILIF) [4] and cross-chain smart contracts [2], have demonstrated the feasibility of connecting disparate blockchain networks to establish unified energy markets. These innovations enhance operational efficiency, support the integration of diverse renewable sources, and contribute to grid stability in decentralized energy systems [6].

This chapter examines the transformative role of blockchain interoperability in reshaping multi-energy systems. It highlights cutting-edge frameworks and real-world case studies, including Power Ledger and Energy Web Chain, to showcase applications and benefits. The remainder of this chapter is structured as follows: Sect. 2 analyzes interoperability challenges, including data silos and cross-chain communication barriers. Section 3 explores blockchain's potential in energy systems, focusing on its features and applications in P2P energy trading. Section 4 delves into interoperability frameworks and case studies, while Sect. 5 outlines the benefits of blockchain interoperability. Finally, Sect. 6 identifies key challenges and proposes future research directions.

2 Problem Analysis: Interoperability Challenges

Blockchain technology offers promising solutions for decentralized energy systems, yet significant interoperability challenges remain. Two major issues are:

- **Data Silos:** Isolated blockchain platforms operate independently with unique protocols, leading to inefficient collaboration and integration of energy resources [1,5].
- **Cross-Chain Barriers:** Challenges like latency, security vulnerabilities, and scalability limit seamless communication between blockchains [2,4]. Relay-chain-based solutions, for example, require high computational resources, making them less viable for large-scale deployment [6].

2.1 Comparison of Interoperability Solutions

Table 1 highlights the key features of two prominent solutions addressing blockchain interoperability challenges in energy systems: BAILIF and Cross-Chain Smart Contracts.

Table 1. Comparison of Interoperability Solutions

Feature	BAILIF	Cross-Chain Smart Contracts
Scalability	High scalability using decentralized notaries	Medium scalability limited by underlying protocols
Security	Robust decentralized attestation protocols	Strong inherent blockchain security features
Ease of Integration	Moderate effort for multi-blockchain integration	Easier adoption leveraging existing platforms
Latency	Real-time transactions with low latency	Low to medium latency based on contract complexity
Energy Efficiency	Efficient validation processes	Dependent on underlying blockchain infrastructure

2.2 Literature Review of Solutions

Recent literature has proposed innovative frameworks to address interoperability challenges. The Blockchain Agnostic Interoperability Framework (BAILIF) uses decentralized notary services and cross-chain attestation protocols to enable communication across diverse blockchains without sacrificing decentralization [4]. Wu et al. [2] explored cross-chain smart contracts, which automate transaction processes and reduce reliance on intermediaries, further facilitating energy distribution in decentralized systems. These solutions highlight the potential for interoperability frameworks to transform energy markets by bridging isolated blockchain platforms and ensuring seamless cross-chain communication.

3 Blockchain as a Solution for Multi-Energy Systems

3.1 Key Blockchain Features

Blockchain technology offers unique features that make it an ideal solution for challenges in multi-energy systems. Its transparency ensures that every transaction is visible to authorized participants, reducing the need for intermediaries and enabling trustless operations [1]. Security is maintained through cryptographic techniques that protect data against unauthorized access or tampering, ensuring the integrity of energy transactions [3]. Additionally, blockchain's immutability guarantees that once a transaction is recorded, it cannot be altered or deleted. This is critical for creating a reliable, auditable record in energy markets, fostering stakeholder confidence and regulatory compliance [7].

3.2 Peer-to-Peer (P2P) Energy Trading

Blockchain facilitates decentralized energy trading by enabling peer-to-peer (P2P) interactions between energy producers and consumers, known as "prosumers." In this model, prosumers can trade surplus energy directly with other users without relying on intermediaries, thereby reducing costs and inefficiencies [4]. For example, platforms like Power Ledger leverage blockchain to record and verify energy trades, ensuring transparency and fairness in transactions [2]. This decentralized approach democratizes energy access and empowers individuals and communities to participate actively in energy markets [4] (Fig. 1).

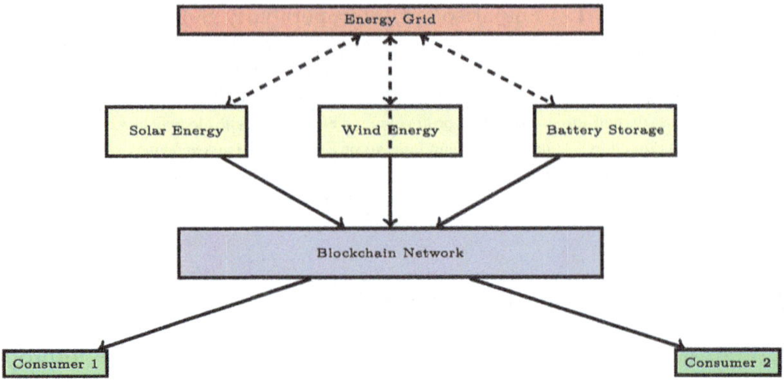

Fig. 1. Blockchain-based peer-to-peer energy trading system.

3.3 Technological Advantages

Blockchain technology provides several advantages for energy systems:

1. Decentralized Architecture: Eliminates single points of failure and ensures robust system operation [2].
2. Enhanced Transparency: Enables stakeholders to trace energy generation, distribution, and usage with accuracy [3].
3. Smart Contracts: Automates energy trading agreements, reducing administrative overhead and transaction delays [1].
4. Scalability: Advanced blockchain platforms like Ethereum and Polkadot are evolving to handle higher transaction volumes, making them suitable for multi-energy systems [4].

By addressing these technological needs, blockchain establishes itself as a foundational tool for creating flexible, secure, and efficient energy ecosystems.

4 Frameworks for Blockchain Interoperability

Blockchain technology has revolutionized energy trading by enabling decentralized and transparent transactions. However, as energy systems increasingly adopt blockchain platforms, interoperability challenges have emerged as a significant barrier to scalability and efficiency. Existing blockchain systems often operate independently, with unique protocols, consensus mechanisms, and data formats, creating "data silos" that inhibit seamless collaboration across networks [1].

Interoperability frameworks aim to address these challenges by enabling communication, asset exchange, and data sharing across heterogeneous blockchain networks. Such frameworks play a crucial role in realizing the full potential of blockchain-enabled energy systems, fostering integration, and driving innovation. This section explores key interoperability frameworks and their applications in multi-energy systems, focusing on the Blockchain Agnostic Interoperability Framework (BAILIF) and real-world platforms like Power Ledger and Energy Web Chain.

4.1 The Blockchain Agnostic Interoperability Framework (BAILIF)

The Blockchain Agnostic Interoperability Framework (BAILIF) is a decentralized protocol designed to overcome interoperability challenges in blockchain networks. At its core, BAILIF utilizes decentralized notary services and cross-chain attestation protocols to facilitate secure data sharing and transactions across heterogeneous blockchain systems. This framework ensures that communication between different blockchain networks is seamless, while maintaining decentralization and transparency [4].

BAILIF operates by using decentralized notaries to validate cross-chain transactions without relying on a central authority. These notaries ensure that data integrity is maintained, enabling trustless interactions across platforms. For instance, a transaction initiated on a private blockchain for solar energy generation can be securely recorded and validated on a public blockchain that tracks wind energy production. This capability addresses one of the primary barriers to interoperability: the lack of a unified communication protocol across blockchain systems [8].

The framework also incorporates cross-chain attestation protocols, which verify the authenticity and consistency of data across networks. These protocols play a critical role in enabling energy trading, as they ensure that records of energy production, distribution, and consumption are synchronized across platforms. By doing so, BAILIF minimizes the risk of discrepancies and fraud, fostering trust among stakeholders in decentralized energy markets [1].

In terms of scalability, BAILIF is designed to handle a high volume of cross-chain transactions without compromising efficiency. Its decentralized nature ensures resilience against single points of failure, making it suitable for large-scale energy systems that integrate diverse energy sources such as solar, wind, and battery storage. However, challenges remain, including the computational overhead required for cross-chain verification and the need for widespread adoption of the framework to achieve its full potential [9] (Fig. 2).

4.2 Case Studies of Platforms

The practical implementation of blockchain interoperability can be observed through platforms like Power Ledger and Energy Web Chain. These platforms demonstrate how blockchain technology can address energy system challenges, enabling efficient, secure, and decentralized energy trading.

4.2.1 Power Ledger

Power Ledger is a blockchain-based platform designed to facilitate peer-to-peer (P2P) energy trading. It allows prosumers-producers and consumers of energy-to trade surplus energy directly with other users without the need for intermediaries. Power Ledger uses blockchain to record energy transactions transparently, ensuring that all trades are secure and verifiable [2].

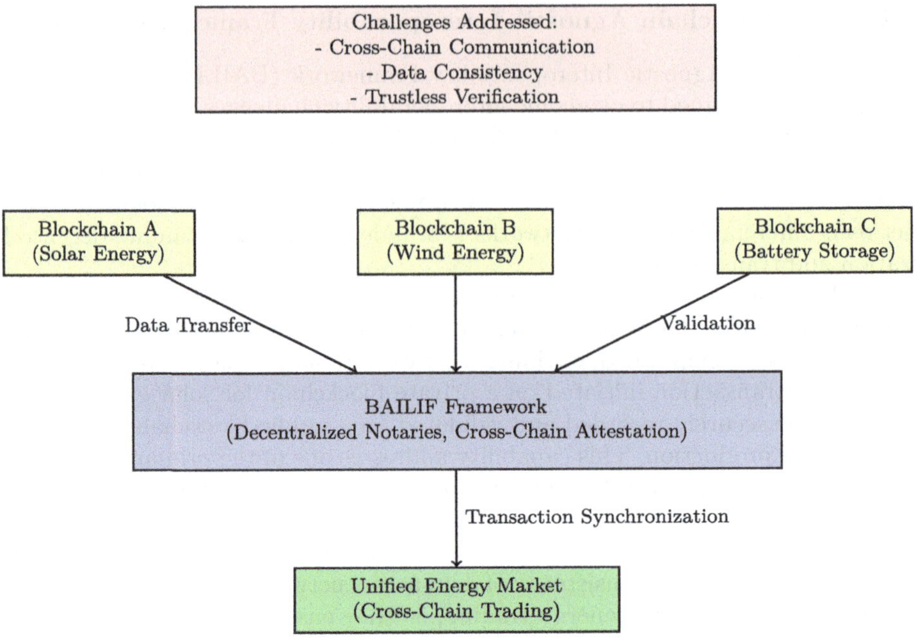

Fig. 2. BAILIF's Operation: Decentralized Notaries and Cross-Chain Attestation Protocols

One of the key features of Power Ledger is its use of smart contracts, which automate the trading process by executing predefined agreements between parties. For instance, if a prosumer generates excess solar energy, the system automatically matches them with a consumer in need of energy, completing the transaction seamlessly. This approach reduces administrative overhead and enhances trading efficiency [4].

Additionally, Power Ledger's integration with local energy grids ensures that energy flow is balanced and optimized, preventing overloads and ensuring grid reliability. By enabling localized energy trading, Power Ledger empowers communities to actively participate in energy markets, promoting sustainability and energy independence [10].

4.2.2 Energy Web Chain

Energy Web Chain is another blockchain-based platform focused on large-scale energy trading and grid management. Unlike Power Ledger, which targets localized energy trading, Energy Web Chain is designed to support grid operators, energy producers, and other large-scale stakeholders in managing energy distribution and transactions across multiple regions [1].

A key innovation of Energy Web Chain is its decentralized operating system (EW-DOS), which provides tools for integrating blockchain into existing energy infrastructure. EW-DOS supports applications such as renewable energy certification, demand response programs, and grid balancing, making it a versatile solution for modern energy systems [11].

Energy Web Chain also emphasizes regulatory compliance, ensuring that its blockchain solutions align with regional energy policies and standards. This focus on compliance makes it a preferred choice for large-scale energy projects that require adherence to strict legal frameworks [3] (Table 2).

Table 2. Comparison of Power Ledger and Energy Web Chain

Feature	Power Ledger	Energy Web Chain
Primary Focus	Peer-to-peer (P2P) energy trading	Large-scale grid management
Key Technology	Smart contracts for automated energy trading	Decentralized Operating System (EW-DOS)
Target Audience	Prosumers and localized energy markets	Grid operators and large-scale energy producers
Grid Integration	Localized energy grid integration	Multi-region grid balancing
Regulatory Compliance	Limited focus on regulations	Strong emphasis on regional policy alignment
Applications	Solar energy trading, community-driven energy solutions	Renewable energy certification, demand response

The table above summarizes the key differences between Power Ledger and Energy Web Chain, highlighting their unique applications and scalability in blockchain-enabled energy systems.

4.3 Use Case of BAILIF in Energy Systems

The Blockchain Agnostic Interoperability Framework (BAILIF) addresses key interoperability challenges by enabling seamless communication and data exchange across heterogeneous blockchains. Using decentralized notaries and cross-chain attestation protocols, BAILIF validates and synchronizes transactions, ensuring trustless interactions across energy platforms. For instance, a solar energy blockchain can seamlessly transfer surplus energy data to a wind energy blockchain, optimizing energy flow and enabling efficient trading.

BAILIF supports three core mechanisms: cross-chain communication for data exchange, decentralized notaries for transaction validation, and attestation protocols for verifying data integrity. These mechanisms eliminate data silos and enhance scalability, making BAILIF suitable for multi-energy microgrids and decentralized energy markets.

Practical applications include energy balancing, where BAILIF enables real-time adjustments between solar, wind, and battery storage systems. Additionally, its support for peer-to-peer energy trading fosters a unified energy market that transcends blockchain-specific boundaries. By enhancing grid resilience and supporting diverse energy sources, BAILIF creates a robust foundation for decentralized energy systems.

However, challenges remain, including scalability, regulatory compliance, and security risks in cross-chain communication. Addressing these limitations will be critical for the widespread adoption of BAILIF and similar frameworks.

4.3.1 Real-World Applications

The practical application of BAILIF can be observed in multi-energy microgrids, where diverse energy sources need to coordinate seamlessly. For example:

- **Energy Balancing**: A microgrid with solar, wind, and battery storage can use BAILIF to optimize energy flow, ensuring that surplus energy from one source can be transferred to areas of demand governed by another blockchain [3].
- **P2P Energy Trading**: BAILIF enables individuals and communities to trade energy across platforms, facilitating a unified energy market that transcends blockchain-specific boundaries [4].
- **Grid Resilience**: By supporting cross-chain transactions, BAILIF enhances grid resilience, enabling energy networks to adapt dynamically to changes in supply and demand [12].

4.3.2 Challenges and Limitations

While BAILIF offers significant advantages, its implementation is not without challenges. Key limitations include:

- **Scalability Issues**: As the number of participating blockchains increases, the computational load on the framework grows, potentially affecting performance.
- **Regulatory Barriers**: The lack of standardized regulations across regions can impede the adoption of cross-chain frameworks like BAILIF [9].
- **Security Risks**: Cross-chain communication introduces potential vulnerabilities, requiring robust cryptographic safeguards to mitigate risks [6].

5 Benefits of Blockchain Interoperability

Blockchain interoperability provides significant advantages for multi-energy systems, addressing critical challenges such as energy allocation, transparency, and the integration of renewable energy sources. Table 3 summarizes the key benefits.

Table 3. Key Benefits of Blockchain Interoperability

Benefit	Description
Enhanced Grid Management	Facilitates real-time data exchange and transaction synchronization, improving energy allocation and grid stability
Transparent and Secure Energy Markets	Eliminates intermediaries, ensuring trust, data integrity, and verifiable transaction records across blockchains
Integration of Renewable Energy Sources	Enables seamless communication between diverse energy sources (e.g., solar, wind, and battery storage) to maximize energy utilization

6 Challenges and Future Directions

Despite the potential benefits, blockchain interoperability in multi-energy systems faces several challenges that need to be addressed to unlock its full potential. This section outlines key issues and explores future research directions.

6.1 Key Challenges

One of the primary challenges is scalability, as current solutions often struggle to manage high transaction volumes in large energy markets. This limitation becomes particularly significant in real-time energy trading, where latency can disrupt the balance between supply and demand. Ensuring low-latency and efficient processing is critical for maintaining the reliability and responsiveness of decentralized energy systems.

Another major concern is the security risks introduced by cross-chain communication. External validators, commonly used in relay-based or notary-based solutions, can become single points of failure, increasing the vulnerability of the entire system. Addressing these risks requires robust security frameworks supported by decentralized attestation protocols and advanced cryptographic techniques to safeguard transaction integrity and prevent fraud.

The fragmented regulatory landscape further complicates the implementation of interoperable systems. Energy policies and data privacy laws vary significantly across regions, making it challenging to design systems that comply with diverse regulations. This lack of standardization hinders the scalability of cross-chain frameworks. Additionally, the high energy consumption associated with blockchain systems, particularly those using Proof-of-Work (PoW) consensus mechanisms, has drawn criticism for its environmental impact. While energy-efficient alternatives like Proof-of-Stake (PoS) offer promise, integrating these into interoperable systems requires further research and development.

6.2 Future Research Directions

Addressing the challenges of blockchain interoperability requires advancements in several areas. Improving scalability and latency is essential, with research focusing on advanced consensus mechanisms and optimized cross-chain protocols to handle high transaction volumes efficiently. Hybrid blockchain models that combine the transparency of public blockchains with the scalability of private ones offer a promising approach to balancing operational needs.

Standardized interoperability protocols are crucial for seamless data exchange, reducing technical barriers and enabling broader adoption of interoperable systems. Artificial intelligence can further enhance efficiency by optimizing cross-chain transactions and predicting energy demand in real time. Additionally, decentralized identity frameworks can strengthen security and data privacy, fostering trust among stakeholders.

By focusing on these areas, blockchain interoperability can evolve into a robust and sustainable framework for multi-energy systems.

7 Conclusion

Blockchain interoperability holds immense promise for transforming multi-energy systems by enabling seamless integration, transparent energy trading, and efficient resource allocation. This chapter has highlighted the key challenges-ranging from scalability and security risks to regulatory and energy efficiency concerns-that must be addressed to fully realize the potential of blockchain in decentralized energy management.

Through the exploration of frameworks like BAILIF and platforms such as Power Ledger and Energy Web Chain, this work has illustrated how cross-chain solutions can overcome interoperability barriers, facilitating collaboration across diverse energy networks. Furthermore, the benefits of interoperability, such as enhanced grid management, secure and transparent markets, and the integration of renewable energy sources, emphasize its critical role in the energy transition.

Future research must prioritize scalable cross-chain communication protocols, hybrid blockchain models, and standardized interoperability frameworks. Integrating AI and decentralized identity systems can further enhance the reliability and efficiency of blockchain applications in energy markets. By addressing these challenges, blockchain technology can pave the way for resilient, adaptive, and sustainable multi-energy ecosystems.

References

1. Zhang, Y., Hou, Z.: Model of decentralized cross-chain energy trading for power systems. Int. J. Eng. Res. **8**, 3614–3628 (2021)
2. Wu, Y., Wu, Y., Cimen, H., Vasquez, J.C., Guerrero, J.M.: P2P energy trading: blockchain-enabled energy society with multi-scale flexibility services. Energy Rep. **8**, 3614–3628 (2022). https://doi.org/10.1016/j.egyr.2022.04.001
3. Taherdoost, H.: Blockchain integration and its impact on renewable energy. Computers **13**, 107 (2024). https://doi.org/10.3390/computers13040107
4. Karumba, S., Jurdak, R., Kanhere, S.S., Sethuvenkatraman, S.: Bailif: a blockchain agnostic interoperability framework. In: Proceedings of the 5th IEEE International Conference on Blockchain and Cryptocurrency (ICBC) (2023). https://doi.org/10.1109/ICBC56567.2023.10174967
5. Saeed, N., Wen, F., Afzal, M.Z.: Decentralized peer-to-peer energy trading in microgrids leveraging blockchain and smart contracts. Energy Rep. **12**, 1753–1764 (2024). https://doi.org/10.1016/j.egyr.2024.01.007
6. Pillai, B., Biswas, K., Muthukkumarasamy, V.: Cross-chain interoperability among blockchain-based systems using transactions. Knowl. Eng. Rev. **35**, 23 (2020). https://doi.org/10.1017/S0269888920000314
7. Mule, N.M., Patil, N., Dhage, O., Kamble, V., Tupe, A.: Decentralizing solar energy transaction with blockchain technology. Int. J. Eng. Res. Technol. (IJERT) **12**(06), 742–748 (2023)
8. Olivieri, L., Spoto, F.: Software verification challenges in the blockchain ecosystem. Int. J. Softw. Tools Technol. Transf. **26**, 431–444 (2024). https://doi.org/10.1007/s10009-024-00758-x

9. Karumba, S.: Blockchain-based decentralized energy systems: advancing scalability, security, privacy, and interoperability. University of New South Wales Thesis (2024). https://doi.org/10.26190/unsworks/28768
10. Chen, S., Ping, J., Yan, Z., Li, J., Huang, Z.: Blockchain in energy systems: values, opportunities, and limitations. Higher Education Press (2022)
11. Bhadange, A., Doshi, R., Karmarkar, T., Shintre, S.: Blockchain-based solution design for energy exchange platform. arXiv (2022). https://doi.org/10.48550/arXiv.2211.13907
12. Chen, S., Guo, B., Yan, H.: Application and prospect of integrated energy interoperability management system based on blockchain. In: 2019 IEEE International Conference on Energy Internet (ICEI), pp. 421–423 (2019). https://doi.org/10.1109/ICEI.2019.00080

A Novel Approach to Liver Disease Diagnosis Classification Using Non-smooth SVM

Z. Khoudi[1](✉), N. Hafidi[1], M. Johri[2], M. Nachaoui[1], and S. Lyaqini[3]

[1] EMI, FST, Sultan Moulay Slimane University, Beni Mellal, Morocco
{zakaria.khoudi,nasreddine.hafidi,m.nachaoui}@usms.ma
[2] ENSAM, Mohammed V University, Rabat, Morocco
m.johri@um5r.ac.ma
[3] LIPIM, Ecole Nationale des Sciences Appliquees, Sultan Moulay Slimane University, Khouribgha, Morocco

Abstract. With two million fatalities each year, liver disease accounts for approximately 4% of all deaths worldwide. Early diagnosis of liver disease is essential for implementing effective control measures and saving lives. The advancement of machine learning has made it possible to predict illnesses and prevent potentially fatal outcomes. Various statistical and machine learning methods have been used to create several prediction models for liver disease. These techniques are applied to datasets to extract valuable insights or build models. In this study, we utilized machine learning methods on the ILPD (Indian Liver Patient Dataset) to develop an intelligent model. Based on experimental results, our proposed approach outperformed seven other supervised machine learning algorithms, achieving the highest accuracy rate of 98.00%.

Keywords: Liver disease · ILP Dataset · Non Smooth SVM · Kernel methods

1 Introduction

Liver disease is a critical and growing global health issue, with conditions ranging from fatty liver disease and cirrhosis to liver cancer. Liver disease occurs when liver cells undergo damage, often resulting from a combination of genetic predispositions, lifestyle factors such as alcohol use, obesity, and infections like hepatitis. According to the World Health Organization (WHO), liver disease affects more than 1.5 billion people worldwide, with liver cancer alone accounting for over 830,000 deaths annually [1]. In a recent study by the Global Burden of Disease (GBD), liver disease cases are projected to increase substantially due to rising obesity rates and other lifestyle factors, placing significant strain on healthcare systems worldwide [2].

N. Hafidi, M. Johri, M. Nachaoui, S. Lyaqini—These authors contributed equally to this work.

A. Laghrib and A. Ghazdali (Eds.): ICMDS 2024, LNNS 1466, pp. 124–132, 2025.
https://doi.org/10.1007/978-3-031-94802-2_12

Early detection and diagnosis of liver disease, particularly liver cancer, can significantly improve treatment outcomes and prevent severe complications. In recent years, advances in machine learning and artificial intelligence (AI) have empowered researchers and clinicians to develop models that can predict liver disease progression and identify high-risk individuals, aiding early intervention efforts [3]. Many machine learning models for liver disease prediction utilize diverse statistical and computational approaches, from regression models to more complex neural networks [4]. AI's role in liver disease management continues to grow due to its robust capabilities in prediction, classification, and clinical assessment [5–9].

Several machine learning algorithms are employed to classify patients with liver disease, with support vector machine (SVM) algorithms being popular. However, in many cases, researchers adopt a square loss function to construct SVM models [10]. Despite its prevalence, the square loss function tends to exaggerate the impact of outliers in classification tasks, rendering prediction algorithms vulnerable [11]. To mitigate the influence of outliers, researchers have turned to hinge loss, which is less sensitive to noise than the square loss function [12,13]. The SVM algorithm is then reformulated as a regularized optimization problem. However, the resulting fidelity term in the minimization problem is non-differentiable. To address this challenge, smooth approximation techniques are employed, transforming the max function into a smoothed variant [14,15]. Subsequently, the modified minimization problem is solved using Adam's algorithm [16]. The paper concludes with numerous numerical validations of the proposed method, comparing it to existing machine learning algorithms. The results underscore the effectiveness of the proposed technique in diagnosing liver disease, outperforming seven supervised machine-learning approaches. This enhanced machine learning model offers a valuable tool for early liver disease detection, confirming and indicating the presence of liver disease.

The paper's structure is organized as follows: Sect. 2 outlines the classification problem's context, transforming it into a minimization problem using the smooth hinge loss function and the optimization approach. Section 3 comprehensively examines comparison experiments, focusing on prediction accuracy, the F-measure metric. Finally, Sect. 4 summarizes the conclusion.

2 Setting of the Problem

For the binary classification in $\mathbb{R}^n$, The training set can be expressed as follows:

$$\mathcal{T} = \{(x_j, y_j)\}_{j=1}^{n}$$

where $\boldsymbol{x}_j \in \mathbb{R}^n$ and $y_j \in \{-1, 1\}$, let $\boldsymbol{\theta} = \left(\boldsymbol{w}^T, b\right)^T$, a SVM model tries to seek a decision hyperplane $f(x) = \boldsymbol{w}^T \Phi(\boldsymbol{x}) + b = 0$, That maximally distinguishes between two types of data. The smooth SVM model may be expressed for the training set as a convex optimization problem with equation constraints in (1)

$$\min_{\boldsymbol{\theta}} C\|\boldsymbol{w}\|^2 + C \sum_{j=1}^{n} L\left(y_j f(x_j)\right) \tag{1}$$

where $\Phi(\boldsymbol{x}_j)$ maps $\boldsymbol{x}_j$ into a higher-dimensiona Y space, $C > 0$ is the regularization parameter [17,18], and $L(s)$ is the hinge loss.

One function that is often used for training classifiers like the SVM is the hinge loss. Here is how the hinge loss function is defined (Fig. 1):

$$L(s) = \max(1 - s, 0).$$

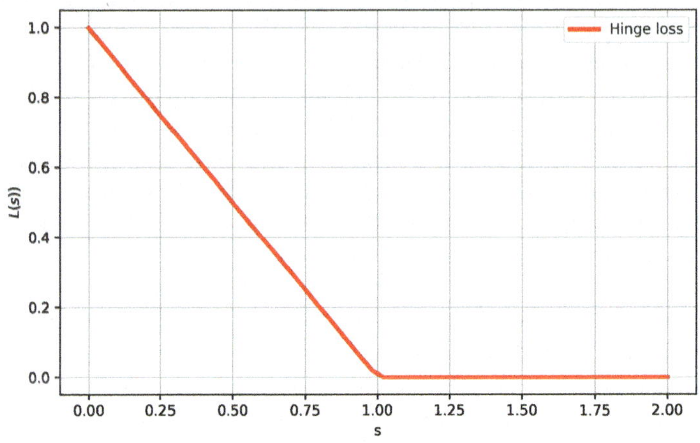

Fig. 1. Hinge loss function

As a result, it cannot be used with optimization techniques that demand that the objective function be differentiable since the hinge loss function is not. Fortunately, there is a workaround for this constraint. The hinge loss function [12,19], described as follows, may be precisely approximated by a smooth. The max function is approximated using a smooth function for every $s \in \mathbb{R}$.

$$\max(s, 0) = \left(s + \frac{1}{\epsilon} \log\left[1 + e^{(-\epsilon s)}\right]\right), \qquad \forall \epsilon > 0.$$

Then the smoothed hinge loss $L_\epsilon(s)$ with parameter ϵ defined by

$$L_\epsilon(s) = \left(1 - s + \frac{1}{\epsilon} \log\left[1 + e^{-\epsilon(1-s)}\right]\right), \qquad \forall \epsilon > 0.$$

As illustrated in the Fig. 2, the smoothed hinge loss L_ϵ approaches the hinge loss L as ϵ goes to $+\infty$.

In the sequel, we consider the approximate optimization problem of (1) given by

$$\min_{\theta} C\|\boldsymbol{w}\|^2 + C \sum_{j=1}^{n} L_\epsilon\left(y_j f(x_j)\right) \tag{2}$$

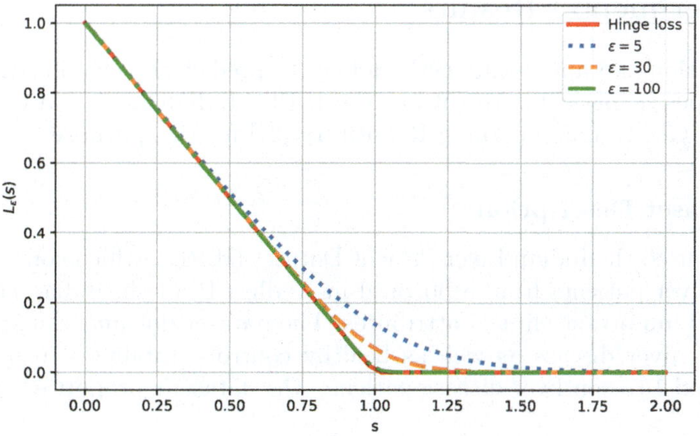

Fig. 2. Smoothed hinge loss

Thus, the objective function in (2) is differentiable.

Problem (2) may be solved using the kernel approach in a larger replicating kernel Hilbert space (RKHS) $\mathcal{K}$ as

$$\min_{f\in\mathcal{K}} C\|\boldsymbol{f}\|_{\mathcal{K}}^2 + \sum_{j=1}^{n} L_\epsilon\left(y_j f(x_j)\right) \tag{3}$$

In RKHS, the Representer Theorem [20] shall be valid, and the solution to the problem of minimizing a regularized risk function can be given as follows:

$$f^* = \sum_{j=1}^{n} \beta_j K\left(\boldsymbol{x}_j, \cdot\right),$$

where K is a kernel function in RKHS and the coefficient $\beta_j \in \mathbb{R}$.

Consequently, using the Representer Theorem in RKHS, we can further express the primal model in problem. (3)) as

$$\min_{\boldsymbol{\beta},b} C\boldsymbol{\beta}^{\mathrm{T}}\boldsymbol{K}\boldsymbol{\beta} + \sum_{j=1}^{n} L_\epsilon\left(\left(y_j, \boldsymbol{\beta}^T\boldsymbol{K}_j + b\right)\right) \tag{4}$$

where K is the Gram matrix derived from associated kernel function $\boldsymbol{K}_{ij} = K\left(\boldsymbol{x}_i, \boldsymbol{x}_j\right)$, $\boldsymbol{K}_j$ represents the i th row of $\boldsymbol{K}$ and $\boldsymbol{\beta}^T = (\beta_1, \beta_2, ..., \beta_n) \in \mathbb{R}^n$.

Various nonlinear optimization techniques may resolve the issue (4). The gradient-type approaches are excellent in resolving this sort of optimization issue. Because of its simplicity and minimal storage requirements, the stochastic Adam method is among the most successful gradient-type techniques (smooth Adam) [16].

3 Experimental Results

In this section, we apply our methodology to predict the classification of liver disease types in patients. We utilize the ILPD (Indian Liver Patient Dataset) from the UCI Machine Learning Repository [21] for this purpose.

3.1 Dataset Description

The study used the Indian Liver Patient Dataset (ILPD), which contains records from 583 liver patients from a hospital in Andhra Pradesh, India. This dataset includes 10 important clinical attributes. The dataset documents patients diagnosed with liver disease as well as healthy controls, totaling 416 liver disease patients and 167 non-liver disease patients, the dataset's properties are detailed in [21].

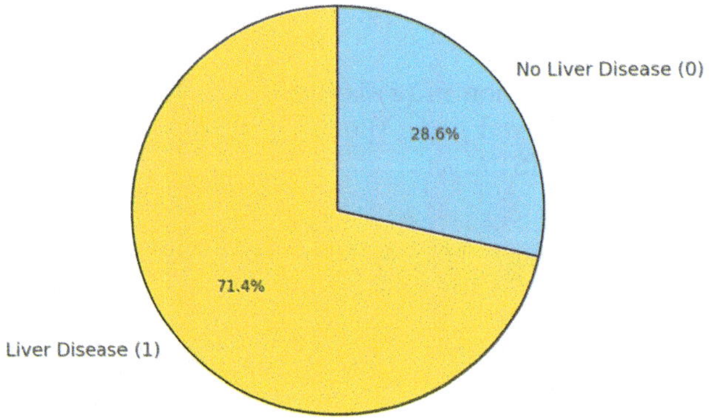

Fig. 3. ILPD Variable response distribution

Figure 3 illustrates the distribution of liver disease in the ILPD dataset, with 71.4% of cases diagnosed with liver disease (label 1) and 28.6% without liver disease (label 0). This distribution indicates that liver disease cases are more prevalent in this dataset, with the response variable serving as the binary target for classification. Specifically, the response variable (or dependent variable) is labeled as "Liver Disease" and takes one of two values:

- **1**: The patient is diagnosed with liver disease.
- **0**: The patient is not diagnosed with liver disease.

This binary classification provides the foundation for the machine learning model's prediction task, aiming to accurately classify patients based on clinical attributes. The dataset's imbalance towards liver disease cases suggests a focus on identifying and distinguishing affected individuals from healthy ones, which is crucial for developing robust diagnostic models.

3.2 Data Splitting and Performance Measures

The dataset was divided into training and testing sets, with an 80% to 20% split, respectively. After training, the model was evaluated using the test dataset, and various performance metrics were recorded. The performance of the proposed model was assessed using popular metrics: Accuracy, Precision, Recall, and F1-score, as defined in Table 1.

Table 1. Performance Metrics

Metric	Formula
Accuracy	$\frac{TP+TN}{TP+TN+FP+FN}$
Precision	$\frac{TP}{TP+FP}$
Recall	$\frac{TP}{TP+FN}$
F1-score	$2 \times \frac{Precision \times Recall}{Precision+Recall}$

where TP = True Positive, TN = True Negative, FP = False Positive, and FN = False Negative.

3.3 Results and Discussions

In this research study, a series of experiments were conducted using a diverse set of machine learning algorithms, including Decision Trees (DT), Random Forests (RF), Logistic Regression (LR), Support Vector Machines (SVM), Naive Bayes (NB), K-Nearest Neighbors (KNN), and XGBoost (XGB) [22–28]. The primary goal of these experiments was to assess the predictive performance of these classification algorithms in identifying the type of liver disease. Various evaluation metrics, such as prediction accuracy and the F-measure, were used to assess their performance.

The outcomes of these experiments are summarized in Table 2, which compares the effectiveness of Decision Trees (DT), Random Forests (RF), Logistic Regression (LR), Support Vector Machines (SVM), Naive Bayes (NB), K-Nearest Neighbors (KNN), and XGBoost (XGB) against our proposed approach. Our approach demonstrates superior performance across all metrics relative to the other algorithms. Specifically, it achieves the highest Accuracy (98.00%), indicating the greatest overall correctness in predictions. The F1-Score, balancing Precision and Recall, is also the highest at 79.00%, reflecting our model's robustness in identifying actual positive cases while minimizing false positives. Furthermore, our approach achieves perfect Recall (77%), effectively capturing all positive cases, with a high Precision of 82.00%, indicating a strong rate of correct positive predictions. These results underscore the strength and reliability of our approach in accurately predicting liver disease types, making it a highly effective tool for this classification task.

Table 2. Performance measures for each algorithm

Model	Accuracy	Precision	Recall	F1-Score
DT	0.72	0.46	0.60	0.52
RF	0.75	0.52	0.43	0.47
LR	0.76	0.56	0.30	0.39
SVM	0.74	0.00	0.00	0.00
NB	0.51	0.34	1.00	0.51
KNN	0.67	0.38	0.50	0.43
XGB	0.72	0.45	0.43	0.44
Our approach	**0.98**	0.82	0.77	0.79

Table 3 presents a comparison with state-of-the-art methods. The results indicate that our method outperforms others, achieving a 98.00% accuracy in early liver disease prediction, surpassing the KNN method by 1.48%. This improvement highlights its superior efficacy in data processing and model optimization, which may contribute to more accurate and earlier diagnoses.

Table 3. Comparison of performance to previous studies' results

Method	[29]	[30]	[31]	[32]	[4]	**Our approach**
Accuracy	96.52	88.40	88.00	93.00	87.48	**98.00**

In summary, the experimental results lead us to the following conclusions regarding the accuracy of our proposed approach: Table 2 shows that our approach outperforms the Decision Trees (DT), Random Forests (RF), Logistic Regression (LR), Support Vector Machines (SVM), Naive Bayes (NB), K-Nearest Neighbors (KNN), and XGBoost (XGB) models. Additionally, the quantitative results in Table 3 demonstrate that the proposed approach surpasses existing state-of-the-art methods. The obtained results confirm that the proposed smooth SVM is an enhanced machine learning model for classification tasks. With the support of this advanced early diagnosis system, diseases can be accurately detected in their initial stages, allowing for the prevention of more severe symptoms.

4 Conclusion

This study introduces a robust classification approach using smooth SVM, optimized with the Adam algorithm, to enhance early liver disease detection. The model is designed to assist medical technicians in diagnosing liver disease effectively. Evaluation results demonstrate its superior performance compared to

Decision Trees (DT), Random Forests (RF), Logistic Regression (LR), traditional Support Vector Machines (SVM), Naive Bayes (NB), K-Nearest Neighbors (KNN), and XGBoost (XGB) models. The proposed approach shows significant potential for liver disease classification. Furthermore, the developed prediction model can be utilized in a mobile application, enabling early disease detection for the general population.

Statements and Declarations

Conflict of Interest. The authors declare that they have no confict of interest.

Availability of Data and Materials. The data used this study are available from https://archive.ics.uci.edu/dataset/225/ilpd+indian+liver+patient+dataset.

References

1. World Health Organization: Hepatitis (2024). https://www.who.int/news-room/fact-sheets/detail/hepatitis. Accessed 09 Nov 2024
2. Younossi, Z.M., Wong, G., Anstee, Q.M., Henry, L.: The global burden of liver disease. Clin. Gastroenterol. Hepatol. **21**(8), 1978–1991 (2023)
3. Lu, F., et al.: Artificial intelligence in liver diseases: recent advances. Adv. Ther. **41**(3), 967–990 (2024)
4. Wu, C.-C., et al.: Prediction of fatty liver disease using machine learning algorithms. Comput. Methods Programs Biomed. **170**, 23–29 (2019)
5. Ahn, J.C., Connell, A., Simonetto, D.A., Hughes, C., Shah, V.H.: Application of artificial intelligence for the diagnosis and treatment of liver diseases. Hepatology **73**(6), 2546–2563 (2021)
6. Khoudi, Z., Nachaoui, M., Lyaqini, S.: Identifying the contextual factors related to the reading performance of Moroccan fourth-grade students from a machine learning-based approach. Educ. Inf. Technol. **29**(3), 3047–3073 (2024)
7. Khoudi, Z., Nachaoui, M., Lyaqini, S.: Finding the contextual impacts on students' mathematical performance using a machine learning-based approach. Infocommun. J. **16**(SI), 12–21 (2024)
8. Hafidi, N., Khoudi, Z., Nachaoui, M., Lyaqini, S.: Enhanced SMS spam classification using machine learning with optimized hyperparameters. Indones. J. Electr. Eng. Comput. Sci. **37**(1), 356–364 (2025)
9. Nachaoui, M., Lyaqini, S., Chaouch, M.: Indicating if water is safe for human consumption using an enhanced machine learning approach. Stat. Optim. Inf. Comput. **11**(1), 70–81 (2023)
10. Ahsan, M.M., Siddique, Z.: Machine learning-based heart disease diagnosis: a systematic literature review. Artif. Intell. Med. 102289 (2022)
11. Zheng, X., Zhang, L., Yan, L.: CTSVM: a robust twin support vector machine with correntropy-induced loss function for binary classification problems. Inf. Sci. **559**, 22–45 (2021)
12. Lyaqini, S., Nachaoui, M.: Identification of genuine from fake banknotes using an enhanced machine learning approach. In: International Conference on Numerical Analysis and Optimization Days, pp. 59–70. Springer (2021)

13. Lyaqini, S., Nachaoui, M., Hadri, A.: An efficient primal-dual method for solving non-smooth machine learning problem. Chaos, Solitons Fractals **155**, 111754 (2022)
14. Lyaqini, S., Hadri, A., Afraites, L.: Non-smooth optimization algorithm to solve the linex soft support vector machine. ISA Trans. **153**, 322–333 (2024)
15. Lyaqini, S., Hadri, A., Ellahyani, A., Nachaoui, M.: Primal dual algorithm for solving the nonsmooth twin SVM. Eng. Appl. Artif. Intell. **128**, 107567 (2024)
16. Haji, S.H., Abdulazeez, A.M.: Comparison of optimization techniques based on gradient descent algorithm: a review. PalArch's J. Archaeol. Egypt/Egyptol. **18**(4), 2715–2743 (2021)
17. Afraites, L., Hadri, A., Laghrib, A., Nachaoui, M.: A non-convex denoising model for impulse and gaussian noise mixture removing using bi-level parameter identification. Inverse Probl. Imaging (2022)
18. Afraites, L., Atlas, A.: Parameters identification in the mathematical model of immune competition cells. J. Inverse Ill-Posed Probl. **23**(4), 323–337 (2015)
19. Lyaqini, S., Nachaoui, M., Quafafou, M.: Non-smooth classification model based on new smoothing technique. In: Journal of Physics: Conference Series, vol. 1743, p. 012025. IOP Publishing (2021)
20. Schölkopf, B., Herbrich, R., Smola, A.J.: A generalized representer theorem. In: International Conference on Computational Learning Theory, pp. 416–426. Springer (2001)
21. Ramana, B., Venkateswarlu, N.: ILPD (Indian Liver Patient Dataset). UCI Machine Learning Repository (2022). https://doi.org/10.24432/C5D02C
22. Kingsford, C., Salzberg, S.L.: What are decision trees? Nat. Biotechnol. **26**(9), 1011–1013 (2008)
23. Rigatti, S.J.: Random forest. J. Insur. Med. **47**(1), 31–39 (2017)
24. Cox, D.R.: The regression analysis of binary sequences. J. Roy. Stat. Soc.: Ser. B (Methodol.) **20**(2), 215–232 (1958)
25. Cortes, C., Vapnik, V.: Support-vector networks. Mach. Learn. **20**(3), 273–297 (1995)
26. Webb, G.I., Keogh, E., Miikkulainen, R.: Naïve bayes. Encyclopedia Mach. Learn. **15**, 713–714 (2010)
27. Kramer, O.: K-nearest neighbors. In: Dimensionality Reduction with Unsupervised Nearest Neighbors, pp. 13–23. Springer (2013)
28. Chen, T., et al.: Xgboost: extreme gradient boosting. R package version 0.4-2 **1**(4), 1–4 (2015)
29. Mitu, M., Mondal, M.N.I., Shovan, S.: Machine learning approach to predict liver disease effectively. In: 2022 4th International Conference on Electrical, Computer & Telecommunication Engineering (ICECTE), pp. 127–130. IEEE (2022)
30. Ghazal, T.M., Rehman, A.U., Saleem, M., Ahmad, M., Ahmad, S., Mehmood, F.: Intelligent model to predict early liver disease using machine learning technique. In: 2022 International Conference on Business Analytics for Technology and Security (ICBATS), pp. 1–5. IEEE (2022)
31. Al Telaq, B.H., Hewahi, N.: Prediction of liver disease using machine learning models with PCA. In: 2021 International Conference on Data Analytics for Business and Industry (ICDABI), pp. 250–254. IEEE (2021)
32. Sivasangari, A., Reddy, B.J.K., Kiran, A., Ajitha, P.: Diagnosis of liver disease using machine learning models. In: 2020 Fourth International Conference on I-SMAC (IoT in Social, Mobile, Analytics and Cloud) (I-SMAC), pp. 627–630. IEEE (2020)

Exploring the Potential of Untrained Neural Networks Based on Optimal Transport for Accelerated MRI

Abdelaadim Khriss[1(✉)], Aissa Kerkour Elmiad[1], Abdelghani Ghazdali[2], and Mohammed Badaoui[3]

[1] Lab. LARI, FSO, Mohammed First University, Oujda, Morocco
abdelaadim.khriss@ump.ac.ma
[2] Lab. LIPIM, ENSAK, Sultan Moulay Slimane University, Khouribga, Morocco
[3] Lab. MSD, ESTO, Mohammed First University, Oujda, Morocco

Abstract. In recent years, deep neural networks have shown great potential in improving the speed and accuracy of medical image reconstruction. However, the practical application of these networks is often limited due to the need for large amounts of data and significant computational resources. Recently, untrained CNNs such as Deep Image Prior and ConvDecoder have produced impressive results in image reconstruction tasks without requiring any training data. In this study, we propose a new approach to improve the performance of an untrained neural network for magnetic resonance imaging (MRI) reconstruction. By incorporating an optimal transport function, we can efficiently transfer the statistical properties of a fully-sampled MRI image to a subsampled image without the need for additional training data. Our experimental results show that this approach outperforms existing state-of-the-art untrained techniques for accelerated MRI reconstruction. Additionally, using the optimal transport function significantly reduces the required computational resources, making it a promising avenue for accelerating MRI reconstruction in clinical settings.

Keywords: MRI Acceleration · Deep Learning · Un-trained CNNs · Optimal Transport · Wasserstein metric

1 Introduction

MRI, or magnetic resonance imaging, is a medical imaging technique that uses magnetic fields and radio waves to produce detailed images of structures inside the body. MRI is a non-invasive medical imaging technique widely used in medicine because it does not involve exposure to ionizing radiation, making it a safer alternative to X-rays and CT imaging [1]. However, one of the drawbacks of MRI is that it is time intensive [2,3]. An MRI scanner creates a strong

A. Kerkour Elmiad, A. Ghazdali and M. Badaoui—These authors contributed equally to this work.

A. Laghrib and A. Ghazdali (Eds.): ICMDS 2024, LNNS 1466, pp. 133–144, 2025.
https://doi.org/10.1007/978-3-031-94802-2_13

magnetic field that aligns protons in body tissue. Radio waves are then used to stimulate the protons, releasing energy as they return to their original configuration. The energy emitted by the protons is detected by a scanner and used to produce images of the body's internal structures. The process of aligning the protons and detecting the emitted energy takes time, which limits the speed of the MRI. Due to these constraints in the scanning process, performing an MRI test can be time consuming. As a result, the research field of compressed sensing emerged, with the goal of accelerating MRI [4] by building an image from few measurements, which is reflected in a faster scan time.

Traditional MRI reconstruction techniques focus on reconstructing high-quality images from subsampled data [5–7]. These methods include regularized image reconstruction methods based on sparsity or low-rank constraints, and several regularization elements can be selected, such as B. 2D wavelet, total variation, dictionary, 3D wavelet, 3D sparse-kt and 3D low-rank [8].

However, traditional compressive sensing methods involve minimizing the ℓ_1 norm or the total variation TV norm without using training data [9]. Although these methods are used in modern MRI scanners, they are being surpassed by a new class of deep learning-based reconstruction techniques, for example, as is demonstrated by the challenge of fast MRI [10], a competition to speed up MRI reconstruction.

In the first part of this paper, we introduce the problem and the current state of research. We then explain our proposed method in detail, including the neural network architecture and the use of optimal transport function. We also describe the data used for testing. Next, we present our experimental results and performance metrics, including comparisons with other state-of-the-art methods. We also provide visualizations and examples of reconstructed images to demonstrate the effectiveness of our method. Finally, we explain the results and discuss the strengths and weaknesses of our proposed method, as well as future work that can be done to further improve its performance.

2 Background

2.1 Problem Statement: MRI Reconstruction

The process of reconstructing high-quality images from subsampled data often involves dealing with an inverse problem. In order to accomplish this, a multi-coil imaging model is utilized. The model is represented by the equation:

$$\mathbf{y} = \mathbf{A}\mathbf{x} + \boldsymbol{\eta} \quad \text{where} \quad \mathbf{A}_i = \mathbf{M}\mathbf{F}_i\mathbf{S}_i \tag{1}$$

In this equation, x represents the image that needs to be reconstructed, while y represents the subsampled measurement data. The noise itself is represented by the variable $\boldsymbol{\eta}$. The coil measurement operator is represented by the variable $\mathbf{A}$, which is made up of three different matrices. The first is the sampling matrix $\mathbf{M}$, which sets some of the frequencies of the k-space measurement to zero. The second is the Fourier transform operator $\mathbf{F}$. Finally, there is the sensitivity map matrix $\mathbf{S_i}$, which is specific to each coil [8].

In the case of accelerated imaging, a subsampling technique is used in k-space, using a mask M that suppresses specific measurement frequencies subsampled by a factor of K speeds up the scanning process by a similar factor. To reconstruct an image from these subsampled measurements, compressed sampling is used to estimate the image from a small number of measurements. In the case of multi-coil imaging, each coil provides an individual measurement. The sensitivity map for each coil is generally unknown but can be estimated from downsampled measurements. Therefore, multi-coil reconstruction poses a significant problem in practice [11].

2.2 Deep Learning Techniques for Enhanced Image Reconstruction

Deep learning has demonstrated significant potential in MRI reconstruction, offering faster and more accurate results compared to traditional methods [10]. Its ability to learn complex mappings between input and output data makes it well-suited for producing high-quality MRI images [12].

Convolutional Neural Networks (CNNs) are widely adopted in medical image reconstruction due to their proficiency in capturing spatial features. Represented as a function $f_\theta : \mathbb{R}^m \rightarrow \mathbb{R}^n$, where θ denotes learnable parameters, CNNs are trained using a dataset $D = \{(x_i, y_i)\}$, with x_i as input images and y_i as corresponding output images. The U-Net architecture is particularly effective, utilizing an encoder-decoder structure to accurately reconstruct MRI images, even in the presence of noise or artifacts [13,14].

Generative Adversarial Networks (GANs) are also employed for MRI reconstruction. They generate new data resembling a given dataset and are defined as $g_\theta : \mathbb{R}^k \rightarrow \mathbb{R}^n$, where θ represents learnable parameters. During training, a dataset $D = \{x_i\}$ is used to approximate the probability distribution of the original data. GAN-based methods, such as those proposed by Tran et al., incorporate cyclic loss functions to enhance consistency between reconstructed and original images, achieving superior image quality compared to traditional compressed sensing methods [15,16].

Untrained neural networks, such as the Deep Image Prior [17] and ConvDecoder [11], have recently garnered attention in MRI reconstruction. These networks autonomously learn underlying patterns without relying on prior training. They use fixed functions $f_\theta : \mathbb{R}^k \rightarrow \mathbb{R}^n$ to map input data, and inversion involves solving $F(f_\theta(z)) = y$. The Deep Image Prior leverages CNN structures to recover images from corrupted versions, while the ConvDecoder reconstructs images from low-dimensional representations through unsupervised training, yielding promising results for enhanced image recovery and diagnostic accuracy.

2.3 Optimal Transport Problem

The optimal transport problem, also known as the Monge-Kantorovich problem [18], seeks to transport a mass distribution from one location to another with minimal cost, typically measured by a distance metric. Monge first proposed a discrete formulation in the 18th century [19], later generalized by Kantorovich in the 20th century to include arbitrary mass distributions [20].

The Monge problem aims to find a transport map $T : X \to Y$ that minimizes the cost:

$$\int_X c(x, T(x))d\mu(x),$$

subject to $T_{\#}\mu = \nu$, where $T_{\#}\mu(B) = \mu(T^{-1}(B))$ for measurable $B \subseteq Y$.

The Kantorovich formulation replaces the transport map with a transport plan $\gamma \in \Pi(\mu, \nu)$, the set of joint probability measures with marginals μ and ν. It minimizes:

$$\int_{X \times Y} c(x, y)d\gamma(x, y),$$

while ensuring $\gamma(A \times Y) = \mu(A)$ and $\gamma(X \times B) = \nu(B)$ for measurable sets $A \subseteq X$ and $B \subseteq Y$.

The **Wasserstein distance**, or Earth-Mover's Distance (EMD) [21], measures the effort required to transform one distribution into another:

$$W(\mu, \nu) = \inf_{\pi \in \Pi(\mu,\nu)} \int_{X \times X} c(x, y)d\pi(x, y),$$

where $c(x, y)$ is the cost function, and π is a joint probability measure with marginals μ and ν.

Applications of optimal transport include its use in Generative Adversarial Networks (GANs), where the Wasserstein distance serves as the objective function in Wasserstein GANs (WGANs) [22]. The Sinkhorn algorithm [23] efficiently approximates transport plans via matrix scaling, significantly reducing computational costs. Optimal transport also addresses domain adaptation [24], aligning distributions between training and testing datasets to enhance model generalization.

In the next section, we detail our methodology, focusing on the ConvDecoder architecture [11], an untrained neural network. We explain its structure and the integration of optimal transport functions to align output predictions with ground truth labels, particularly when input and output distributions differ. This approach enhances the accuracy and reliability of predictions.

3 Methodology

3.1 Untrained Neural Networks for Image Reconstruction

In image reconstruction, un-trained neural networks can be used as image priors, which can be represented mathematically as follows:

Let $G : \mathbb{R}^p \to \mathbb{R}^{c \times w \times h}$ be a convolutional image prior that maps a parameter space of size p to an image space of size $c \times w \times h$, where c is the number of channels, and w and h are the width and height of the image, respectively. For example, for a grayscale image, $c = 1$.

Deep-learning-based image models are typically trained to learn a parameterized function that maps an input to an output. The output of such models is an image, and the weight parameters are trained on a distribution of images. This can be represented mathematically as follows:

$$M_\theta(x) = y$$

where M_θ is the trained image model parameterized by the weight parameters θ, x is the input to the model, and y is the output image.

In contrast, un-trained neural networks are image models where the input to the network is relatively irrelevant and fixed, and the weights are the parameters of the model. The output of the model is again an image, which can be represented mathematically as follows:

$$G_\theta(z) = y$$

where G_θ is the un-trained neural network parameterized by the weight parameters θ, z is a fixed input to the model, and y is the output image.

3.2 Optimizing Multi-coil MRI Reconstruction with Optimal Transport

To reconstruct Multi-coil accelerated MRI using sensitivity map estimates, we assume that the multi-coil acquisition is done with a coil sensitivity map that is known or estimated beforehand. The sensitivity map vector $S \in \mathbb{C}^M$ is defined, where each element corresponds to the sensitivity of a particular coil. We also define the operator S that maps an image to its sensitivity-weighted image in k-space.

To estimate the image based on the measurements y, we use a Multi-coil accelerated MRI reconstruction algorithm that incorporates sensitivity map estimates. Specifically, we define the sensitivity map vector $S \in \mathbb{C}^M$, where each element corresponds to the sensitivity of a particular coil. We also define the operator S that maps an image to its sensitivity-weighted image in k-space.

We then solve the following optimization problem:

$$L(C) = \inf_{\gamma \in \Gamma(y, MFS(C))} \int d(MFS(C), y) d\gamma(MFS(C), y)$$

where C is the parameter vector of a neural network $G : \mathbb{R}^p \to \mathbb{R}^{c \times w \times h}$, and MFS is the Fourier transform of the sensitivity-weighted image. The use of the optimal transport Wasserstein distance as the loss function helps to incorporate the spatially varying sensitivity of the coils and can lead to improved image quality compared to other loss functions.

We use an iterative algorithm to minimize the objective function $L(C)$, alternating between updating the sensitivity-weighted image and the sensitivity map estimate. The reconstructed image can then be obtained by applying an inverse Fourier transform to the sensitivity-weighted image and taking the absolute value.

This method provides a way to reconstruct high-quality images from undersampled k-space measurements while taking into account the spatially varying sensitivity of the coils.

3.3 Enforcing Data Consistency

To ensure faithful image reconstruction, data consistency is enforced by incorporating missing frequency information from subsampled measurements. Frequencies from specific coordinates on the x-axis, $S_x = \{i_1, \ldots, i_m\}$, are extracted from the original image and used to fill gaps in the Fourier domain. These components replace corresponding subsampled frequencies, preserving essential features and aligning the reconstruction with acquired data. This method reduces the required frequencies, optimizing resources while maintaining critical information and enhancing the neural network's reconstruction of missing regions.

3.4 Network Architectures

This study utilizes the *ConvDecoder* architecture [11], a neural network designed for image generation, which maps a low-dimensional parameter space z to an output image x as:

$$x = G(z), \tag{2}$$

where z has dimensions $c_0 \times w_0 \times h_0$ and the output image x has dimensions $c \times w \times h$.

The ConvDecoder consists of multiple layers with up-sampling, convolution, ReLU activation, and Batch Normalization operations. Up-sampling increases the resolution by a factor of 2, convolution extracts local pixel information, and ReLU introduces non-linearity:

$$R(x) = \max(0, x). \tag{3}$$

Batch Normalization normalizes each channel independently, improving stability and convergence:

$$B(i)(x) = \left(\frac{x - \text{mean}(x)}{\sqrt{\text{var}(x) + \epsilon}} \right) \times \gamma(i) + \beta(i), \tag{4}$$

where $\gamma(i)$ and $\beta(i)$ are learnable scale and shift parameters.

The final layer combines the channels using a 1×1 convolutional layer:

$$F(x) = \text{conv}(x, w(L)) + b(L), \tag{5}$$

producing the reconstructed image $x(L)$.

The next section presents experimental results, including performance metrics, comparisons with other untrained methods, and visualization of reconstructed images, demonstrating the effectiveness of this approach.

4 Experimental Results

4.1 Dataset

Our study focused on reconstructing multi-coil knee images because multi-coil is clinically more relevant than single-coil reconstruction, we employed the recently released fastMRI dataset [25] and followed the same approach as described in ConvDecoder [11]. The dataset comprises two sets, a training and a validation set, consisting of complete k-space measurements of knees captured with 15 coils, and brains captured with varying numbers of coils. Additionally, the dataset includes "reference" images, which are generated by reconstructing the coil images from each coil measurement as $\hat{x}_i = F^{-1}y_i$ and then combining the coil images using the root-sum-of-squares (RSS) algorithm, resulting in a final single image:

$$\hat{x} = \sqrt{\sum_{i=1}^{nc} |\hat{x}_i|^2} \tag{6}$$

In the equation above, $|\cdot|$ and $\sqrt{\cdot}$ denote element-wise absolute value and squared root operations, respectively. As the coil sensitivities overlap only minimally, the RSS algorithm is effective in merging the images (Fig. 1).

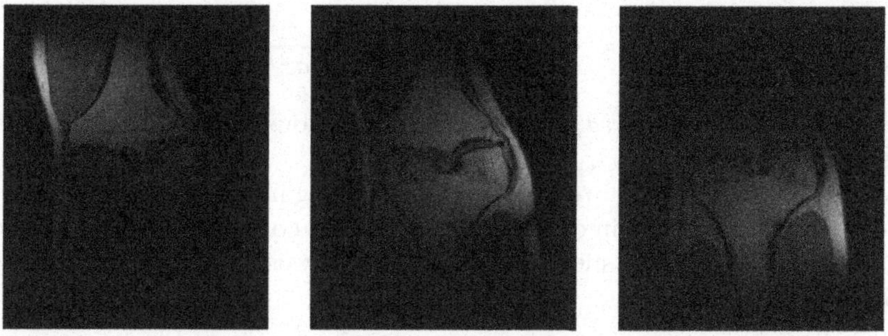

Fig. 1. Spatial images of individual coils from fully sampled data [25].

Our focus was on accelerated imaging, which involved obtaining measurements with a mask. To achieve this, we employed a standard 1D variable density mask 4×, which includes random or equi-spaced vertical lines across the k-space. This type of mask is challenging yet practically relevant and is the default option in the fastMRI challenge. During the evaluation, we compared the generated images with the "reference" images reconstructed from the full k-space (Fig. 2).

4.2 Performance Measures

Peak signal-to-noise ratio (PSNR) and structural similarity index (SSIM) are widely used metrics for evaluating reconstructed images compared to ground

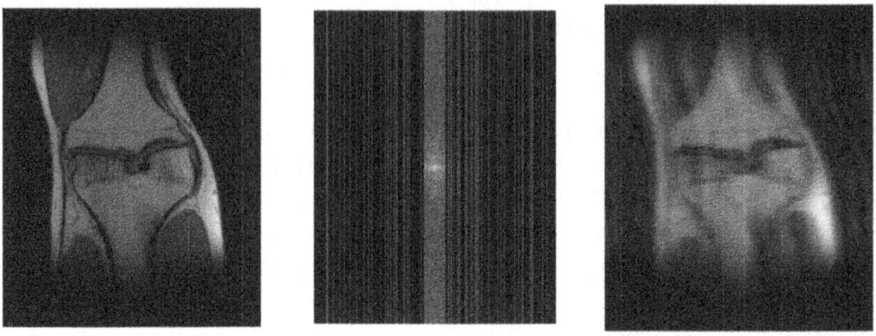

Fig. 2. Standard 1D variable density mask 4× [25].

truth [26]. PSNR quantifies noise levels in the reconstruction, with higher values indicating less noise, and is calculated as:

$$PSNR = 20 \cdot \log_{10}\left(\frac{MAX}{\sqrt{MSE}}\right), \tag{7}$$

where MAX is the maximum pixel value and MSE is the mean squared error.

SSIM measures structural, luminance, and contrast similarity, ranging from −1 to 1, with 1 indicating identical images. It is defined as:

$$SSIM = \frac{(2\mu_x\mu_y + C_1)(2\sigma_{xy} + C_2)}{(\mu_x^2 + \mu_y^2 + C_1)(\sigma_x^2 + \sigma_y^2 + C_2)}, \tag{8}$$

where μ_x, μ_y are means, σ_x, σ_y are standard deviations, and σ_{xy} is the covariance of the images.

While PSNR and SSIM are effective for assessing image quality, they may not fully capture performance in cases of high contrast, complex textures, or visually critical details. Combining metrics and visual analysis is often recommended for comprehensive evaluation.

4.3 Evaluation Results

4.3.1 Performance

We evaluated the performance of three untrained methods, namely ConvDecoder, Deep Image Prior and our method based on the optimal transport function. Our study was carried out using accelerated multi-coil knee measurements from the fastMRI dataset.

From Table 1 The proposed method based on the optimal transport function produced remarkable results, with an SSIM value of 0.8304 and a PSNR of 31.57. Compared to other methods such as ConvDecoder and Deep Image Prior [27], it outperformed in terms of reconstructed image fidelity. The improved results obtained with optimal transport are mainly attributed to its ability to preserve the properties of the original image while efficiently transporting structural properties (Fig. 3).

Table 1. Average results obtained through MRI reconstruction.

Method	SSIM	PSNR
Proposed Method	**0.8304**	**31.57**
ConvDecoder	0.8127	31.30
DIP	0.7649	28.16

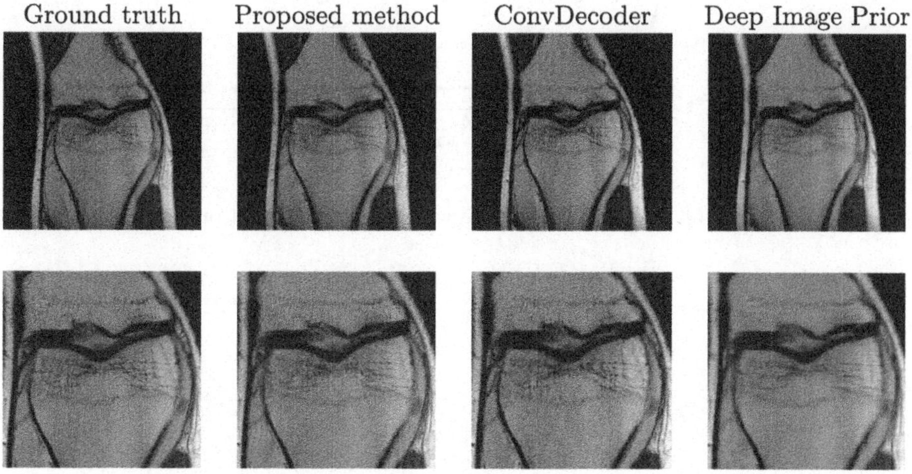

Fig. 3. Examples of reconstructions for the proposed method, ConvDecoder and Deep Image Prior. The top line represents an extended version of the bottom line.

4.3.2 Convergence Analysis of Optimal Transport Loss Function

The values of the loss function decrease progressively with each iteration, suggesting convergence of the algorithm. From the Fig. 4 The first iterations show a greater decrease in the loss function, indicating a rapid improvement in image reconstruction. However, as the algorithm progresses, the decrease in the loss function becomes slower, suggesting that image reconstruction reaches a plateau where further improvements are more subtle. The values of the optimal transport loss function obtained at the end of the algorithm are relatively low, indicating a good match between source and reconstructed image pixels. These preliminary results suggest that using the optimal transport loss function may be promising. Further studies will be needed to compare the performance of this loss function with others to assess the visual quality of the reconstructed image.

The results Fig. 5 show a progressive increase in SSIM and PSNR values with each iteration. Initially, the values are very low, indicating low similarity between the original image and its first reconstruction. However, as iterations progress, SSIM and PSNR values increase steadily.

This increase suggests that the iterative image reconstruction process is improving as new information is used to refine the reconstruction. Specifically,

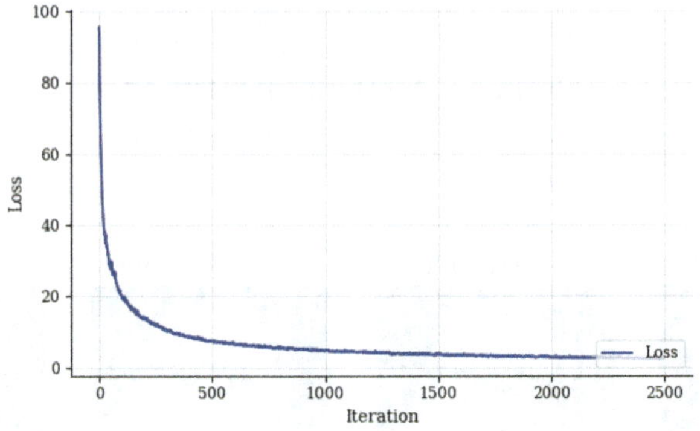

Fig. 4. The convergence rate of the loss function.

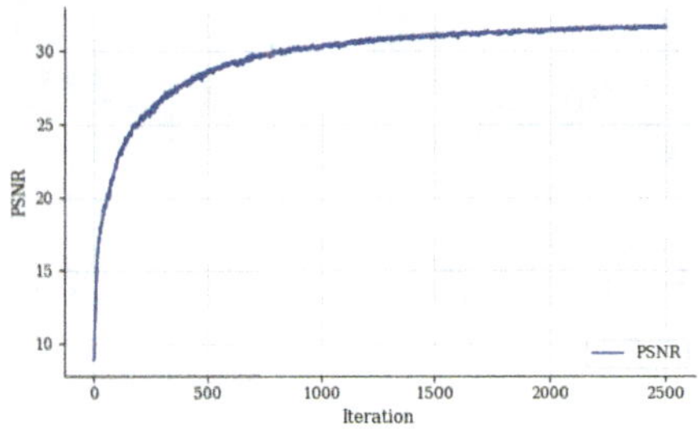

(a) PSNR

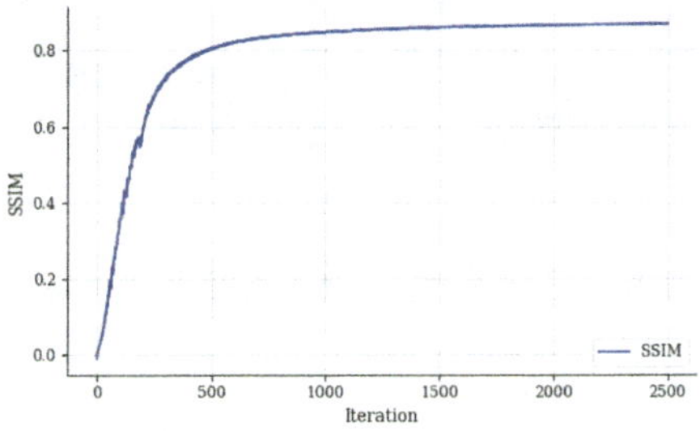

(b) SSIM

Fig. 5. Iterative Image Reconstruction: Analyzing SSIM and PSNR Improvement

with each iteration, additional features are added to the reconstruction, leading to a better match with the original image. This can be attributed to the accumulation of information and the reduction of reconstruction errors.

As the number of iterations increases, SSIM and PSNR continues to increase. This suggests that the reconstruction becomes increasingly accurate and that the structure of the original image is better captured.

However, it's important to note that there may be a limit to the improvement in similarity. Once the iterations reach a certain point, the gains in similarity may decrease or level off, indicating that the reconstruction has reached an optimal level of accuracy given the information available.

5 Conclusion

This study presents a novel untrained neural network approach using optimal transport for accelerated MRI reconstruction. The proposed method outperforms state-of-the-art untrained techniques, achieving superior SSIM and PSNR metrics while efficiently transferring the statistical properties of fully-sampled MRI images to subsampled ones without requiring additional training data. The parameter grid search ensures optimal performance and reduces computational demands, making this method a promising candidate for clinical MRI reconstruction. Future directions include applying the method to broader datasets, comparing it to advanced models, and incorporating additional features to improve its effectiveness in medical imaging applications.

References

1. Radiological Society of North America (RSNA): MRI Safety (2021). Accessed 5 Apr 2023
2. Hennig, J.: Multiecho imaging sequences with low refocusing flip angles. J. Magn. Reson. **1969**(78), 397–407 (1988)
3. El Malki, I., El Hakoume, A., Laghrib, A., Hadri, A.: Improved image denoising via self-supervised weickert operator learning and plug-and-play learned primal dual. Neurocomputing **620**, 129267 (2025)
4. Lustig, M., Donoho, D.L., Santos, J.M., Pauly, J.M.: Compressed sensing MRI. IEEE Signal Process. Mag. **25**(2), 72–82 (2008)
5. Ghazdali, A., Hadri, A., Laghrib, A., Nachaoui, M.: Poisson noise and gaussian noise separation through copula theory. Multimedia Tools Appl. **83**(26), 67927–67952 (2024)
6. Laghrib, A., Bella, F., Nachaoui, M., Hakim, M.: Learning nonlocal weights for second-order nonlocal super-resolution. Discrete Continuous Dyn. Syst.-S **18**(1), 238–268 (2025)
7. Ghazdali, A., Ourdou, A., Hakim, M., Laghrib, A., Mamouni, N., Metrane, A.: Robust approach for blind separation of noisy mixtures of independent and dependent sources. Appl. Comput. Harmon. Anal. **60**, 426–445 (2022)
8. Wang, T., Zhang, G., Li, W., Li, S., Guo, D., Zhang, X.: A review on deep learning MRI reconstruction without fully sampled k-space. Signal Process. Image Commun. **97**, 117978 (2021)
9. Block, K.T., Uecker, M., Frahm, J.: Undersampled radial MRI with multiple coils. Iterative image reconstruction using a total variation constraint. Magn. Reson. Med. **57**(6), 1086–1098 (2007)
10. Knoll, F., et al.: Advancing machine learning for MR image reconstruction with an open competition: overview of the 2019 fastmri challenge. Magn. Reson. Med. (2020)
11. Darestani, M.Z., Heckel, R.: Accelerated MRI with un-trained neural networks. IEEE Trans. Comput. Imaging **7**, 724–733 (2021). https://doi.org/10.1109/TCI.2021.3097596
12. Hammernik, K., et al.: A review and experimental evaluation of deep learning methods for MRI reconstruction. Magn. Reson. Med. **79**(6), 3040–3057 (2018)
13. Li, Z., Liu, F., Yang, W., Peng, S., Zhou, J.: A survey of convolutional neural networks: analysis, applications, and prospects. IEEE Trans. Neural Netw. Learn. Syst. **33**(12), 6999–7019 (2022). https://doi.org/10.1109/TNNLS.2021.3084827
14. Ronneberger, O., Fischer, P., Brox, T.: U-net: convolutional networks for biomedical image segmentation. CoRR abs/1505.04597 (2015). arxiv:1505.04597

15. Creswell, A., White, T., Dumoulin, V., Arulkumaran, K., Sengupta, B., Bharath, A.A.: Generative adversarial networks: an overview. IEEE Signal Process. Mag. **35**(1), 53–65 (2018). https://doi.org/10.1109/MSP.2017.2765202
16. Quan, T.M., Nguyen-Duc, T., Jeong, W.-K.: Compressed sensing MRI reconstruction using a generative adversarial network with a cyclic loss. IEEE Trans. Med. Imaging **37**(6), 1488–1497 (2018)
17. Ulyanov, D., Vedaldi, A., Lempitsky, V.S.: Deep image prior. CoRR abs/1711.10925 (2017). arxiv:1711.10925
18. Peyré, G., Cuturi, M.: Computational Optimal Transport (2020)
19. Monge, G.: Mémoire sur la Théorie des Déblais Et des remblais. De l'Imprimerie Royale (1781)
20. Kantorovich, L.: On the translocation of masses. J. Math. Sci. **133** (2006). https://doi.org/10.1007/s10958-006-0049-2
21. Panaretos, V.M., Zemel, Y.: Statistical aspects of Wasserstein distances. Ann. Rev. Stat. Appl. **6**(1), 405–431 (2019)
22. Salimans, T., Zhang, H., Radford, A., Metaxas, D.: Improving GANs using optimal transport. arXiv preprint arXiv:1803.05573 (2018)
23. Cuturi, M.: Sinkhorn distances: lightspeed computation of optimal transport. In: Advances in Neural Information Processing Systems, vol. 26 (2013)
24. Flamary, R., Courty, N., Tuia, D., Rakotomamonjy, A.: Optimal transport for domain adaptation. IEEE Trans. Pattern Anal. Mach. Intell **1** (2016)
25. Zbontar, J., et al.: fastMRI: an open dataset and benchmarks for accelerated MRI. arXiv preprint arXiv:1811.08839 (2018)
26. Hore, A., Ziou, D.: Image quality metrics: PSNR vs. SSIM. In: 2010 20th International Conference on Pattern Recognition, pp. 2366–2369. IEEE (2010)
27. Ulyanov, D., Vedaldi, A., Lempitsky, V.: Deep image prior. In: Proceedings of the IEEE Conference on Computer Vision and Pattern Recognition, pp. 9446–9454 (2018)

OP-DCNN: Optimized Deep Convolutional Neural Network for Facial Expression Recognition

Maryam Knouzi(✉), Fatima Zohra Ennaji, and Imad Hafidi

Laboratory of Process Engineering, Computer Science and Mathematics (LIPIM), National School of Applied Science-USMS (ENSA-KH), Khouribga, Morocco
maryam.knouzi@usms.ac.ma, {f.ennaji,i.hafidi}@usms.ma

Abstract. Attitudes, gestures, postures, facial expressions, and speech are considered to be channels for the transmission of human emotions. Visual interaction is a powerful communication tool for individuals. Even a simple spontaneous change in facial expression conveys feelings such as happiness, sadness, surprise, or anxiety. This present work proposes a solution that automatically recognizes the emotion shown on a given face. Thus, a solution based on an optimized deep convolutional neural network (OP-DCNN) is used to classify the following emotions: Happiness, Sadness, Anger, Disgust, Surprise, Neutrality, and Fear. The impact of optimization techniques like regularization and hyperparameter tuning are discussed in this work. It outperforms state-of-the-art results, with an accuracy of 98% on the CK+ dataset.

Keywords: Facial expressions · OP-DCNN · regularization · hypermarameter tuning

1 Introduction

Facial expression recognition is one of the important aspects of computer vision, which focuses on automatically recognizing human emotions through facial expression analysis. It is an essential component of nonverbal communication, the expressions of a person provide information about his intentions, emotions [1], and responses. A wide range of applications, including virtual reality, human-computer interaction, mental health evaluation, and surveillance systems [2], can benefit from accurately identifying these emotions.

The advancement of Deep Convolutional Neural Networks (CNNs) in recent years has greatly improved FER (Facial Expression Recognition) system performance. CNNs are especially efficient at interpreting facial expressions because of their exceptional ability to extract complicated patterns and hierarchical parts from images [3,4]. CNN-based models achieved state-of-the-art accuracy in emotion classification tasks through using their capacity to directly learn discriminative features from raw pixel data. However, there are a number of difficulties in training deep CNNs for FER, including the need for massive labeled

A. Laghrib and A. Ghazdali (Eds.): ICMDS 2024, LNNS 1466, pp. 145–154, 2025.
https://doi.org/10.1007/978-3-031-94802-2_14

datasets, overfitting, and high computational costs. Optimization techniques are necessary for improving the efficiency and performance of deep CNN models in order to overcome these difficulties [5–7]. To improve model accuracy, and minimize overfitting, techniques like hyperparameter tuning, regularization (such as dropout), and efficient optimizers (like Adam and Nadam) are crucial. Additionally, CNNs are optimized through the use of data augmentation, transfer learning, and learning rate scheduling, which improves their capacity to generalize to new and unseen data. The integration of optimization techniques with deep learning and computer vision techniques in Facial Expression Recognition (FER) not only improves model robustness but also reduces training time, making these systems more practical for real-world deployment. Thus, this paper presents the impact of various optimization methods on enhancing the performance of CNN-based FER systems, in order to develop more efficient and accurate models for emotion recognition.

The paper is organized as follows. Related works that implement deep learning methods for facial expression recognition are addressed in Sect. 2. The proposed approach is described in Sect. 3, and the dataset used is provided in Sect. 4. In contrast, Sect. 5 introduces and explains the optimized deep convolutional neural network. Section 6 discusses the experimental study, data evaluation, and key conclusions; Sect. 7 summarizes the paper and presents future research.

2 Related Works

Computers can now recognize patterns in data thanks to machine learning [8]. At the same time, computer vision uses machine learning (ML) to comprehend and analyze visual data, making tasks like facial recognition and image classification possible.

Classification is the final stage of the FER system in which the classifier categorizes the expression as happy, sad, surprised, angry, fear, disgusted and neutral.

In this section, we discuss the different classification techniques that are used to classify the facial expressions.

CNN is the most widely used architecture in computer vision techniques and machine learning. Massive data is required for training in order to fully exploit its ability to solve complex functions.

Mollahosseini et al. [9] proposed a deep CNN network for FER across several available datasets. After extracting facial landmarks from the data, the images were reduced to 48×48 pixels. Next, they applied the data augmentation technique. They presented the possibility of using the network-in-network technique, which increases local performance through locally applied convolution layers, and also reduces over-fitting. They obtained an accuracy of 93.2% for ck+. Lopes et al. [10]studied the impact of data pre-processing before to network training in order to achieve better emotion classification. Using the CK+, JAFFE, BU-3DFE databases, the accuracy for CK+ was 96.76%.

Furthermore, to classify six basic emotions, K. Jain [11] applied a deep convolutional neural network, which is a combination of a convolutional neural network coupled with deep residual blocks. They trained their model on two datasets, CK+ and JAFFE [12], and achieved an accuracy of 95.24% for the CK+ dataset. G. Wen [13] implemented an ensemble approach to overcome the limitations of the CNN using CK+, the accuracy was 76.05%, while Ravi et al.[14] used a CNN for emotion recognition using the CK+, JAFFE, RAFD and YaleFace datasets, they had as accuracy 89.62% for CK+, and 73.81% for JAFFE.

K.Shan [15] used face detection and equalization histogram with 7 classes using the CK+ and JAFFE datasets, they were able to achieve an accuracy of 80.30% for CK+ and 76.74% for JAFFE. In addition, Khorrami et al.[16] demonstrated how an unbiased CNN achieves recognition accuracy using data augmentation, they processed the CK+ and TED datasets, and achieved 95.1% accuracy using 6 classes.

Veena Mayya et al. [17] proposed a deep convolution neural network to recognize the emotions, they achieved an accuracy of 96.02% using CK+, while D.Y. Liliana [18] applied in the CNN fully-connected layers, dropout, as a regularization method using the CK+ dataset, they achieved an accuracy of 92.81%. A concept of Reversible Neural Network is introduced by A. Barman et al.[19], they identified facial expressions using a set of salient features to classify the emotions into six basic categories. They achieved an accuracy of 96.8%.

3 The Proposed Work

This section describes the architecture of the proposed DCNN model.

In this work, we propose feature extraction and classification of 7 basic emotions: anger, happiness, disgust, fear, sadness, surprise, and neutrality based on an Optimized Deep Convolutional Neural Network (OP-DCNN).

Using the CK+ dataset [20], images are pre-processed by cropping using haar cascade for face detection [21,22], resizing to 200×200, grayscale conversion, improving the contrast of the image and applying the data augmentation technique. In this work, we explored several optimization techniques to enhance the performance and efficiency of our deep Convolutional Neural Network (DCNN) model for Facial Expression Recognition (FER). Specifically, we integrated regularization methods such as dropout and batch normalization to prevent overfitting and stabilize training. Additionally, we utilized learning rate reduction and early stopping techniques to accelerate convergence, ensuring the robustness of the model on validation data. The Fig. 1 below illustrates the architecture adopted for this study.

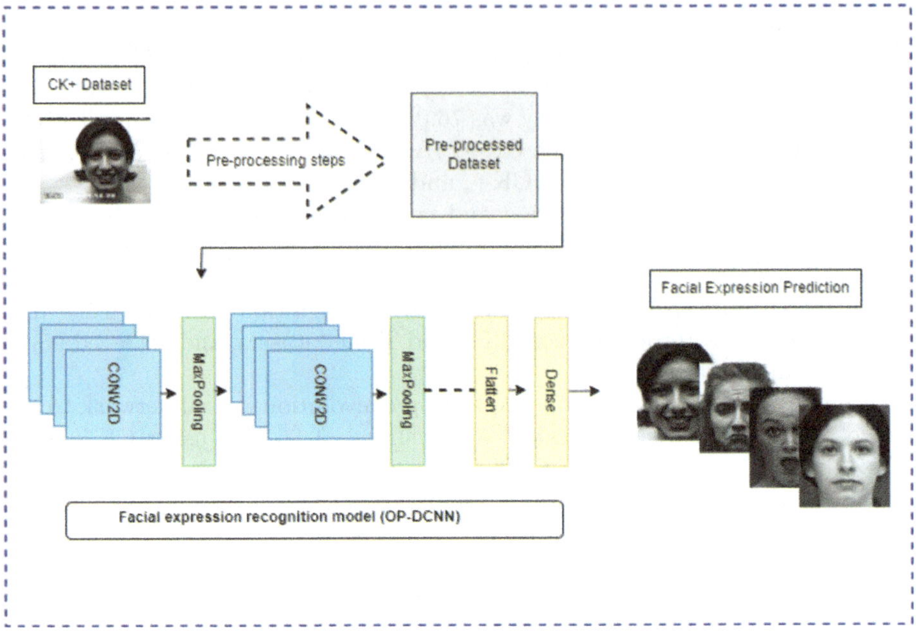

Fig. 1. The architecture of the proposed Optimized Deep Convolution Neural Network

4 Dataset Used

There are several publicly available datasets used for Facial Expression Recognition (FER), each designed to capture a wide range of emotions under varying

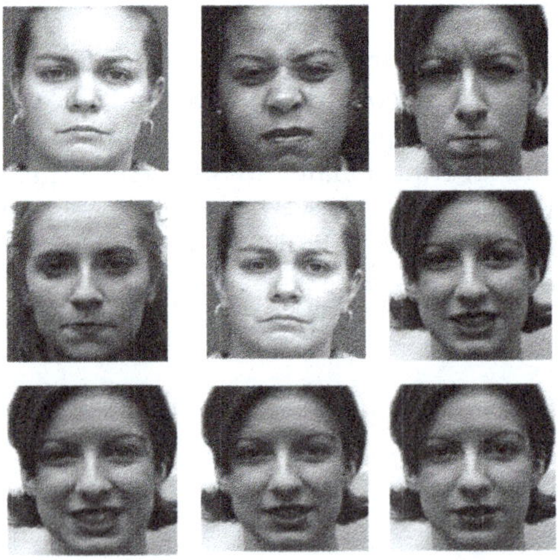

Fig. 2. Sample Images from the CK+ Dataset Showing Different Facial Expressions

conditions such as CK+ [20], Fer2013 [23] and JAFFE[12]. These datasets play a crucial role in training and evaluating the performance of FER models. The Cohn-Kanade extended facial expression dataset is a public dataset for emotion recognition [20]. It includes posed and unposed (spontaneous) expressions. The CK+ dataset includes a total of 593 sequences from 123 people. Sequences can vary from 10 to 50 frames, while images show the transition from a neutral face to a specific expression. This dataset is widely used for facial expression recognition systems. The following Fig. 2 is showing an example of CK+ images.

5 The Optimized Deep Convolutional Neural Network

Convolutional Neural Networks (CNNs) are a specialized form of deep neural networks, generally utilized for image classification tasks. They are designed to automatically learn and extract essential features from input images, which are then used to predict the class or category of an image. A CNN is composed of several interconnected layers, including convolutional layers, pooling layers, and fully connected layers, which work together to analyze and interpret visual data from the input images. The proposed OP-DCNN contains 6 convolution layers. We used the activation function **Elu** or exponential linear unit, this is a function that tends to converge the cost to zero faster and produce more accurate results. Unlike other activation functions, Elu has an additional alpha constant which must be a positive number, it is represented by the following equation:

$$f(x) = x \quad \text{si} \quad x > 0 \tag{1}$$

$$\alpha(e^x - 1) \quad \text{si} \quad x <= 0 \tag{2}$$

The model was initially implemented with images of size **200** × **200** as input to the model. After the first convolution layer, the model features batch normalization. Training deep neural networks with multiple layers is a challenge, as they can be sensitive to initial random weights and the configuration of the learning algorithm, **Batch normalization** is a technique for training very deep neural networks that normalizes the inputs of a layer for each mini-batch, this stabilizes the learning process and considerably reduces the number of training epochs required to train deep networks.

After each hidden layer, an exclusion layer, **Dropout** was inserted, with the exclusion value set to **0.4, 0.4, 0.5 and 0.6** respectively. It randomly deactivates **40%, 50% and 60%** of the hidden layer nodes to avoid overfitting. We used the **Softmax** activation function in the output layer.

$$Softmax(x) = \frac{e^j}{\sum_i e^i} \tag{3}$$

As model optimizer, we chose **Adam** (Adaptive Moment Estimation), with a learning rate of **0.001**, it attempts to calculate adaptive learning rates for each parameter. This is very useful in complex network structures, as different parts of the network have different sensitivities to weight adjustment, it is represented as follows.

$$m_t = \beta_1 m_{t-1} + (1 - \beta_1) g_t \tag{4}$$

$$v_t = \beta_2 v_{t-1} + (1 - \beta_1) g_t^2 \tag{5}$$

where t represents the t-th iteration and g_t is the calculated gradient, m_t and v_t are the moving averages of the gradient and the squared gradient. Next, we chose **Categorical Crossentropy** as the loss function, and the callbacks that were included in the model are **EarlyStopping**, and **ReduceLROnPlateau**, the callbacks are for optimizing the training process of neural networks, especially in preventing overfitting and ensuring efficient learning.

6 Results and Discussion

6.1 Experimental Configuration and Implementation

All programs were run on a DELL vPro laptop equipped with an 8th-generation Intel i5 processor, 16 GB RAM and a 512 GB SSD hard disk.

The programs were developed in the Python programming language. The entire model is simulated in the Jupyter laboratory, the libraries required for these experiments are Keras, Tensorflow, numpy, OpenCV, Sklearn, Seaborn. OpenCV was mainly used for image pre-processing, using the cascade classifier, the model detected the face in the image and cropped it, then it resized the images 200×200, and converted the rgb image into a gray image and enhacing its contrast.

Data augmentation was performed using the keras API. Data augmentation was used for training set images to train the model on a larger number of images and to train the model for invariances on small transformations. These parameters are illustrated in the following Table 1.

Table 1. Data augmentation parameters

Operation Type	Value
Rotation	25
Zoom	0.15
Shear	0.15
Width Shift	0.15
Height Shift	0.15
Horizontal Flip	True

6.2 Results

We present the performance of the proposed model (OP-DCNN) on the CK+ dataset. Our proposed model achieved an accuracy of **98%**, in the following, results and classification metrics are presented (Table 2).

Table 2. Performance Metrics for Facial Expression Recognition

Class	Precision	Recall	F1-Score	Support
Angry	0.99	0.95	0.97	172
Neutral	0.92	1.00	0.96	48
Disgust	0.99	0.98	0.98	182
Fear	1.00	0.93	0.96	112
Happy	0.99	0.99	0.99	261
Sadness	0.97	0.99	0.98	112
Surprise	0.86	1.00	0.93	51

The classification report reveals strong performance across most classes, with particularly high precision, recall, and F1-scores for "happy", "disgust", and "sadness". These classes achieved nearly perfect results, indicating that the model is effective at predicting these emotions. "Fear" also shows good metrics, with perfect precision but slightly lower recall, suggesting it might miss

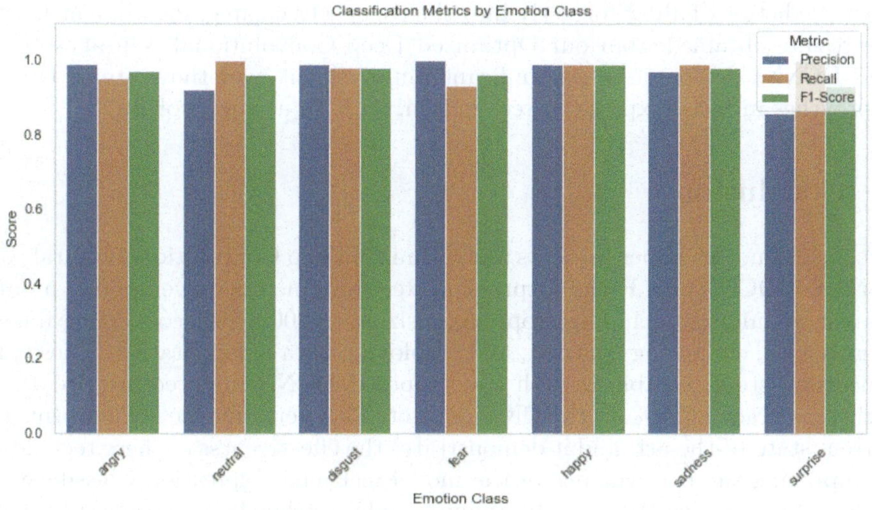

Fig. 3. Classification Metrics by Emotion Class

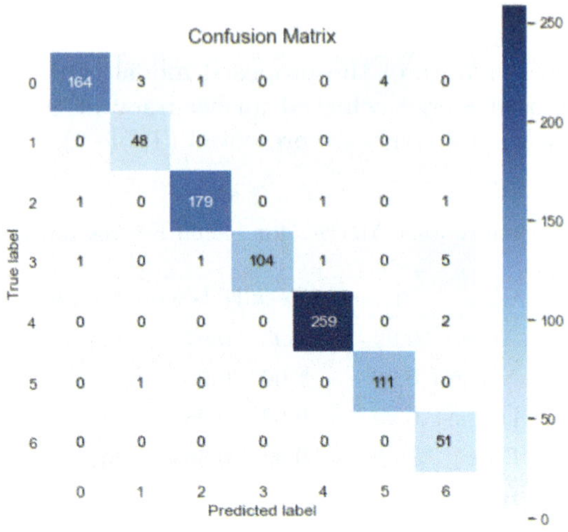

Fig. 4. Confusion Matrix for the proposed OP-DCNN

a small percentage of fear instances. "Neutral" stands out with perfect recall (1.00), meaning the model identifies all neutral instances, but its precision is somewhat lower (0.92). The following Fig. 3 represents the classification metrics for every class (Fig. 4).

The confusion matrix of OP-DCNN model provides detailed information of the performance of the model across each of the seven emotion classes. The model achieved a strong accuracy for the "happy" and "surprise" classes, with most predictions falling correctly into their respective categories. In conclusion, the results obtained from our Optimized Deep Convolutional Neural Network (OP-DCNN) demonstrate a significant improvement over the state of the art approaches in facial expression recognition, with an accuracy of **98%**.

7 Conclusion

In conclusion, this paper presents an Optimized Deep Convolutional Neural Network (OP-DCNN) for Facial Expression Recognition, where we applied preprocessing techniques, including cropping, resizing to 200×200, converting images to grayscale, enhancing contrast, and employing data augmentation. These preprocessing steps, combined with the proposed DCNN architecture, led to an higher accuracy of 98% on the CK+ dataset. This performance outperforms the current state-of-the-art, and it demonstrates the effectiveness of these techniques in improving the performance of the model in Facial Expression Classification. Moving forward, we aim to enhance our model further by exploring additional training techniques, and using multiple datasets.

References

1. Jbel, M., Jabrane, M., Hafidi, I., Metrane, A.: Sentiment analysis dataset in Moroccan dialect: bridging the gap between Arabic and Latin scripted dialect. Lang. Resour. Eval. (2024). https://doi.org/10.1007/s10579-024-09764-6
2. Vyas, A.S., Prajapati, H.B., Dabhi, V.K.: Survey on face expression recognition using CNN. In: 2019 5th International Conference on Advanced Computing and Communication Systems (ICACCS), pp. 102–106 (2019). https://doi.org/10.1109/ICACCS.2019.8728330
3. El Malki, I., El Hakoume, A., Laghrib, A., Hadri, A.: Improved image denoising via self-supervised weickert operator learning and plug-and-play learned primal dual. Neurocomputing **620**, 129267 (2025)
4. Limami, F.-E., Hadri, A., Laghrib, A., Afraites, L.: Fractional optimal control for deep convolutional neural networks exploring ode-based solutions for image denoising. Inverse Probl. Imaging **19**(2), 424–455 (2025)
5. Laghrib, A., Bella, F., Nachaoui, M., Hakim, M.: Learning nonlocal weights for second-order nonlocal super-resolution. Discrete Continuous Dyn. Syst.-S **18**(1), 238–268 (2025)
6. El Malki, I., Jauberteau, F., Laghrib, A., Nachaoui, M.: Enhancing deblurring through optimal controlled bilevel with nonlocal weighted model. Math. Control Related Fields **15**(1), 296–322 (2025)
7. El Hakoume, A., Laghrib, A., Afraites, L., Hadri, A.: A variational pdnet network using a learning reaction-diffusion equation. Expert Syst. Appl. **249**, 123605 (2024)
8. Jabrane, M., Tabbaa, H., Hadri, A., Hafidi, I.: Enhancing entity resolution with a hybrid active machine learning framework: strategies for optimal learning in sparse datasets. Inf. Syst. **125**, 102410 (2024). https://doi.org/10.1016/j.is.2024.102410
9. Mollahosseini, A., Chan, D., Mahoor, M.H.: Going deeper in facial expression recognition using deep neural networks. In: 2016 IEEE Winter Conference on Applications of Computer Vision (WACV), pp. 1–10. IEEE (2016). https://doi.org/10.1109/wacv.2016.7477450
10. Lopes, A.T., de Aguiar, E., De Souza, A.F., Oliveira-Santos, T.: Facial expression recognition with convolutional neural networks: coping with few data and the training sample order. Pattern Recognit. **61**, 610–628 (2017). https://doi.org/10.1016/j.patcog.2016.07.026
11. Jain, D.K., Shamsolmoali, P., Sehdev, P.: Extended deep neural network for facial emotion recognition. Pattern Recogn. Lett. **120**, 69–74 (2019)
12. Zheng, W., Zhou, X., Zou, C., Zhao, L.: Facial expression recognition using kernel canonical correlation analysis (KCCA). IEEE Trans. Neural Networks **17**(1), 233–238 (2006). https://doi.org/10.1109/TNN.2005.860849
13. Wen, G., Hou, Z., Li, H., Li, D., Jiang, L., Xun, E.: Ensemble of deep neural networks with probability-based fusion for facial expression recognition. Cogn. Comput. **9**(5), 597–610 (2017). https://doi.org/10.1007/s12559-017-9472-6
14. Ravi, R., Yadhukrishna, S.V., prithviraj, R.: A face expression recognition using CNN & LBP. In: 2020 Fourth International Conference on Computing Methodologies and Communication (ICCMC), pp. 684–689 (2020). https://doi.org/10.1109/ICCMC48092.2020.ICCMC-000127
15. Shan, K., Guo, J., You, W., Lu, D., Bie, R.: Automatic facial expression recognition based on a deep convolutional-neural-network structure. In: 2017 IEEE 15th International Conference on Software Engineering Research, Management and Applications (SERA), pp. 123–128 (2017). https://doi.org/10.1109/SERA.2017.7965717

16. Khorrami, P., Paine, T.L., Huang, T.S.: Do Deep Neural Networks Learn Facial Action Units When Doing Expression Recognition? (2017). https://arxiv.org/abs/1510.02969
17. Mayya, V., Pai, R.M., Manohara Pai, M.M.: Automatic facial expression recognition using DCNN. Procedia Comput. Sci. **93**, 453–461 (2016). https://doi.org/10.1016/j.procs.2016.07.233. Proceedings of the 6th International Conference on Advances in Computing and Communications
18. Liliana, D.Y.: Emotion recognition from facial expression using deep convolutional neural network. J. Phys. Conf. Ser. **1193**(1), 012004 (2019). https://doi.org/10.1088/1742-6596/1193/1/012004
19. Barman, A., Dutta, P.: Facial expression recognition using reversible neural network. Appl. Soft Comput. **162**, 111815 (2024). https://doi.org/10.1016/j.asoc.2024.111815
20. Lucey, P., Cohn, J.F., Kanade, T., Saragih, J., Ambadar, Z., Matthews, I.: The extended cohn-kanade dataset (CK+): a complete dataset for action unit and emotion-specified expression. In: 2010 IEEE Computer Society Conference on Computer Vision and Pattern Recognition - Workshops, pp. 94–101 (2010). https://doi.org/10.1109/CVPRW.2010.5543262
21. Salmam, F.Z., Madani, A., Kissi, M.: Facial expression recognition using decision trees. In: 2016 13th International Conference on Computer Graphics, Imaging and Visualization (CGiV), pp. 125–130 (2016). https://doi.org/10.1109/CGiV.2016.33
22. Cossetin, M.J., Nievola, J.C., Koerich, A.L.: Facial expression recognition using a pairwise feature selection and classification approach. In: 2016 International Joint Conference on Neural Networks (IJCNN), pp. 5149–5155 (2016). https://doi.org/10.1109/IJCNN.2016.7727879
23. Goodfellow, I.J., et al.: Challenges in representation learning: a report on three machine learning contests. Neural Netw. **64**, 59–63 (2015). https://doi.org/10.1016/j.neunet.2014.09.005. Special Issue on "Deep Learning of Representations"

Blockchain-Powered Artificial Intelligence for Healthcare Systems Data Orchestration

Ayoub Louja[1](✉), Abdellah Jamali[1], and Najib Naja[2]

[1] Hassan First University of Settat, Faculty of Sciences and Technologies Research Laboratory IT, Networks, Mobility and Modeling, Casablanca, Morocco
oulouja.ayoub@gmail.com

[2] The National Institute of Posts and Telecommunications, Rabat, Morocco

Abstract. The artificial intelligence and blockchain technology presents a transformative opportunity for modern healthcare systems. This integration addresses critical challenges such as data security, privacy concerns, effective data exchange, patient confidentiality, and the inherent inefficiencies of centralized infrastructures. By combining the decentralized, immutable, and transparent structure of blockchain with the advanced analytics of AI, this synergy provides a powerful framework for secure, large-scale healthcare data management. Smart contracts, deployed on platforms like Ethereum, facilitate secure, automated transactions of sensitive medical data, reducing human error and enhancing overall reliability. This approach enhances privacy, improves data interoperability, ensures accountability, and optimizes overall system performance. This chapter investigates the multifaceted applications of AI and blockchain in healthcare, particularly focusing on data integrity and more fluid data-sharing protocols. It also delves into the key technical challenges, research directions, and the latest advancements shaping an integrated AI-blockchain ecosystem for healthcare data orchestration. Ultimately, the chapter demonstrates how the synergy between AI and blockchain can revolutionize healthcare data management, leading to more secure, efficient, and transparent healthcare systems.

Keywords: Blockchain · Artificial Intelligence · DAI · Deep Learning · Machine Learning · Smart Contracts · Data Orchestration · Electronic Medical Records · Healthcare

1 Introduction

Blockchain technology functions as a decentralized, distributed ledger composed of cryptographically linked blocks [1]. It has gained considerable traction in various sectors, including the Internet of Things (IoT) [2], decentralized finance (DeFi) [3], and healthcare [4]. In the healthcare context, smart contracts, self-executing programs that eliminate the need for intermediaries [5], have significantly enhanced transparency and efficiency, reducing costs and accelerating workflows.

A. Laghrib and A. Ghazdali (Eds.): ICMDS 2024, LNNS 1466, pp. 155–163, 2025.
https://doi.org/10.1007/978-3-031-94802-2_15

These smart contracts can also automate critical processes such as medical procedure validation, thereby strengthening the security and performance of the Internet of Medical Things (IoMT) [6]. This advancement allows patients to access reliable information about their symptoms more quickly, facilitating remote monitoring and treatment in real time. Additionally, analyzing a patient's environment allows for data processing that can predict their health status.

Meanwhile, AI continues to expand within healthcare, contributing to a wide array of applications, from advanced diagnostics to workflow optimization [7]. However, the reliance on centralized AI systems raises major concerns about data security, system failures, and opaque decision-making algorithms, all of which can undermine patient safety and trust. To overcome these vulnerabilities, integrating AI with blockchain, a concept referred to as decentralized AI (DAI) [8], provides a more secure, transparent, and tamper-resistant framework for sensitive medical information.

Despite these advancements, the healthcare industry faces several critical challenges affecting its efficiency and sustainability. These include securely managing data, ensuring the integrity of pharmaceutical products, enhancing the effectiveness of clinical trials, and preventing fraud. Balancing the protection of electronic health records (EHRs) with the need for accessibility is crucial, especially given the constant threats of data breaches [9]. Establishing a unified platform for sharing medical data across different systems is essential to enable information exchange without compromising sensitive health data [10]. The authenticity of healthcare data also demands innovative solutions to maintain trust without relying on a single governing authority [11].

Pharmaceutical safety is equally vital, necessitating solutions that secure medications throughout their supply chain [12]. Advanced analytic methods, including machine learning and deep learning, are required to handle and interpret increasingly complex patient data[13–15]. The health sector also strives to reduce administrative overheads, maintain quality, accurately forecast demand, automate document organization for regulatory purposes, and improve supply-chain and inventory management [16–18].

Furthermore, the industry faces pressures to reduce administrative costs, improve manufacturing quality and efficiency, forecast demand more accurately, automate document classification for compliance, and enhance supply chain and inventory management [16, 17]. The emerging synergy between AI and blockchain holds the potential to revolutionize the healthcare industry by improving the security, transparency, and orchestration of AI models and data management, ultimately promoting better patient care and data protection. This discussion also covers the ethical implications and future directions for developing a more efficient, transparent, and secure healthcare infrastructure, with the ultimate goal of optimizing patient care and strengthening data protection.

2 Background

2.1 Blockchain in Healthcare

The transition from physical documents to EHRs was intended to address issues of damage and intentional alteration inherent in paper records. However, EHRs remain vulnerable to unauthorized access, data breaches, and permanent loss. The existing security infrastructure is insufficient to fully protect patient health information, raising serious concerns about privacy and data protection [19].

Blockchain technology has emerged as a versatile and transformative tool with significant potential across various sectors, notably in healthcare [20,21]. Its unique features, decentralization, transparency, and immutability, address many longstanding issues in healthcare [22]. Decentralization enables the deployment of distributed applications without reliance on a central authority. The replication of information across all network nodes fosters transparency, allowing healthcare stakeholders, especially patients, to monitor how their data are accessed and used, thereby enhancing trust in the system [20,23].

The immutability property of blockchain ensures that once a record is added, it cannot be altered or deleted, preserving the integrity and accuracy of patient information. Cryptographic algorithms encrypt stored data, ensuring that only authorized individuals with appropriate access permissions can decrypt it, thus strengthening security and privacy. Patient identities are protected through pseudonymization, utilizing cryptographic keys instead of personal identifiers, which facilitates data sharing while maintaining anonymity [19,21,23].

In smart healthcare, which leverages technologies like IoT devices to enhance patient care, the integration of robust security protocols is challenging due to resource constraints and reliance on public networks, increasing vulnerability [23]. Blockchain assists by providing a secure platform for storing and sharing patient data among various healthcare providers. This enhances coordination in managing chronic illnesses and ensures data integrity and transparency [20].

However, to fully realize its potential, especially when integrated with AI, certain challenges must be addressed. Balancing blockchain's inherent transparency with patient confidentiality is crucial, as even anonymized data can sometimes be analyzed to reveal sensitive information. Additionally, issues related to scalability and speed are important; blockchain networks may experience latency, affecting real-time healthcare applications. For instance, a blockchain node validating hospital records in real time could slow down if too many transactions are simultaneously processed, potentially delaying urgent medical decisions (Fig. 1).

2.2 Artificial Intelligence in Healthcare

AI is a broad field focused on developing intelligent agents that process environmental inputs to optimize specific objectives, enabling machines to analyze vast healthcare data, recognize disease patterns, make precise diagnoses, and support advanced decision-making based on large datasets [24]. AI systems learn by training on datasets filled with complex information, which they interpret through iterative processes involving mathematical reasoning [25].

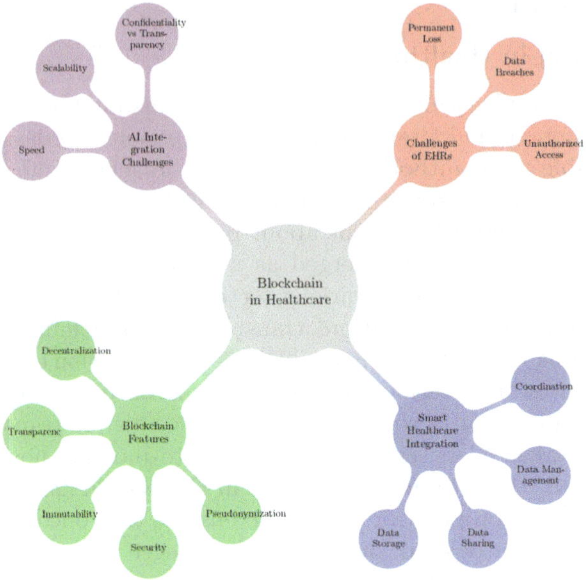

Fig. 1. Blockchain in Healthcare

In healthcare, AI applications range from analyzing extensive data to understand human conditions and recognize disease patterns, to making highly accurate diagnoses and delivering precision health interventions. AI technologies are already employed in various products and services.

In personalized medicine, AI is used for diverse functions: analyzing voluminous datasets, spotting clinical anomalies, aiding in diagnostic accuracy, and supporting data-backed decisions [26].

Building upon these advancements, AI is significantly enhancing patient monitoring, diagnostics, and personalized care through various innovative applications. A notable development is the integration of AI with wearable biosensors and the IoMT. These wearable devices, applied directly to the skin, collect real-time data on individuals' daily activities and vital signs, providing valuable health insights [27]. Companies like Apple have implemented AI-based systems in smartwatches to detect conditions such as atrial fibrillation, offering immediate alerts when physiological parameters deviate from normal ranges [28].

Furthermore, AI is revolutionizing radiology and medical imaging by enhancing the analysis and interpretation of complex imaging data. Traditional medical image analysis is labor-intensive and subjective, heavily relying on the expertise of clinicians. AI algorithms, particularly machine learning and deep learning techniques, such as convolutional neural networks, assist in processing data from modalities like CT scans, MRIs, ultrasounds, and pathology images [28].

AI supports clinical decision-making by analyzing extensive datasets, combining new data with existing records, and providing actionable insights for treatment accuracy [29]. During health crises, AI supports efficient data management by tracking patient trends, anticipating resource needs, and ensuring data accuracy.

Furthermore, AI enhances diagnostic capabilities, including imaging diagnostics and genomic analysis, by leveraging layered data structures that autonomously identify complex patterns. NLP processes health records to assist in disease prediction and treatment planning, while advanced image analysis helps evaluate biological samples for early disease detection. AI-driven wearable technologies monitor patients in real time, providing early warning signals for health complications. Additionally, digital pathology converts slides into digital formats for comprehensive examination. Predictive analytics driven by AI offer preventive diagnostics by assessing disease risks based on genetics, lifestyle, and medical history, making AI an invaluable tool in modern healthcare for personalized, accurate, and timely interventions [30] (Fig. 2).

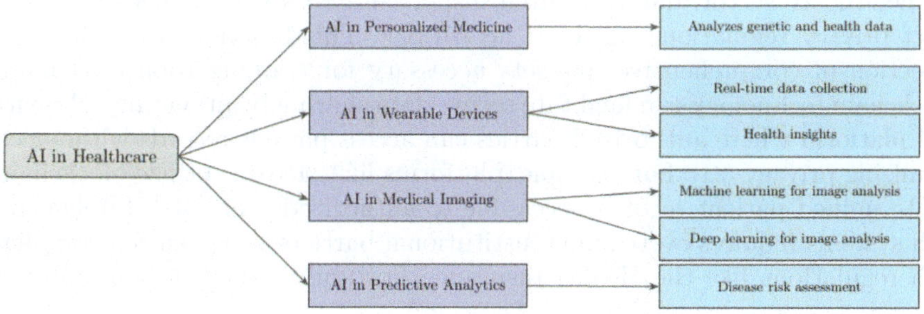

Fig. 2. Artificial Intelligence in Healthcare

2.3 Unified AI–Blockchain Systems for Healthcare

The integration of blockchain and AI offers transformative solutions for healthcare by enhancing data security, improving patient care, and streamlining key processes. This synergy enables innovative approaches in managing EHRs and remote healthcare using virtual and augmented reality. Platforms like MediLinker and Aider illustrate blockchain's role in securing data integrity [31], with MediLinker focusing on EHR connectivity and identity management, and Aider supporting global remote healthcare services. The potential of a healthcare metaverse, combining AI-driven disease diagnostics with blockchain-secured interactions [32], promises a future of secure e-health. In one proposed framework, an AI-assisted smartwatch application is used to detect eye-rubbing events in patients with keratoconus by collecting motion data from the gyroscope and accelerometer[33]. Blockchain integration would store these records immutably, ensuring data provenance and tamper resistance while providing secure access to healthcare providers for proactive intervention and patient education. Ultimately, this integration establishes a strong foundation for advancing healthcare delivery, research, and management.

3 Data Challenges

While the convergence of AI and blockchain technologies holds significant promise, several data-related challenges must be addressed to fully realize this potential. These challenges encompass data availability and accessibility, data quality and integrity, data privacy and security, interoperability and standardization, transparency and trust, and scalability and performance.

3.1 Data Availability and Accessibility

AI systems rely on large and diverse datasets to function effectively [24]. In the healthcare sector, data are often dispersed across various institutions, and strict privacy regulations can limit access [19]. This fragmentation hinders the collection of comprehensive datasets necessary for training robust AI models. Blockchain technology can facilitate secure data sharing by providing a decentralized platform where authorized parties can access patient records without compromising privacy [21]. For example, platforms like MedRec utilize blockchain to create unified patient records accessible to authorized users [34]. Implementing such systems requires overcoming institutional barriers and ensuring compliance with regulations like the Health Insurance Portability and Accountability Act [35].

3.2 Data Quality and Integrity

The effectiveness of AI models is heavily dependent on the quality of the data used for training [25]. Healthcare data can be incomplete or inconsistent due to manual entry errors and a lack of standardized data collection practices. Blockchain's immutability can enhance data integrity by preventing unauthorized alterations [19]. Smart contracts can automate data validation processes, but ensuring accuracy at the point of data entry remains a significant challenge [5].

3.3 Data Privacy and Security

Protecting patient privacy is paramount in healthcare. AI systems often require access to sensitive health information, raising concerns about data breaches and unauthorized access [9]. Blockchain enhances security through cryptographic techniques and decentralized storage, reducing the risk of single points of failure [19]. However, the transparent nature of blockchain may conflict with privacy requirements, as transaction data can be visible to network participants [20]. Techniques like data anonymization, pseudonymization, and the use of private or permissioned blockchains can help address these concerns [10].

3.4 Interoperability and Standardization

Achieving seamless data exchange requires interoperability between healthcare systems and blockchain platforms. The lack of standardized data formats and protocols, along with concerns over data ownership, access control, and organizational trust, poses significant barriers to widespread adoption and integration [22]. Blockchain can promote interoperability by enforcing common standards through smart contracts and shared protocols [20]. Integrating initiatives like the Fast Healthcare Interoperability Resources (FHIR) standards with blockchain and AI systems remains a complex task.

3.5 Transparency and Trust

The opaque nature of some AI algorithms can make it difficult for clinicians and patients to understand how decisions are made, potentially hindering trust and adoption. Blockchain can enhance transparency by providing immutable records of data transactions and AI model updates [36]. Research into explainable AI aims to make AI systems more interpretable without sacrificing performance [37].

4 Conclusion and Future Work

The integration of artificial intelligence with blockchain technology, particularly through decentralized AI, has the potential to revolutionize healthcare data management and decision-making. By combining AI's advanced data analytics with blockchain's decentralized and immutable ledger, this synergy effectively addresses many of the privacy and security challenges in current healthcare systems. The implementation of smart contracts further enhances efficiency by automating essential processes like patient consent and secure data sharing, enabling decisions that meet the latency demands of healthcare settings.

However, despite these promising benefits, substantial challenges remain, especially in terms of scalability and seamless integration with existing healthcare infrastructure.

Future research should aim to refine both blockchain and AI frameworks for deployment in large-scale healthcare environments and develop low-latency consensus algorithms to support real-time decision-making. With these advancements, the fusion of AI and blockchain could become fundamental to next-generation healthcare data management, facilitating rapid, secure, and effective patient care.

References

1. Ibrahimy, M.M., Norta, A., Normak, P.: Blockchain-based governance models supporting corruption-transparency: a systematic literature review. Blockchain Res. Appl. **5**(2), 100186 (2023)
2. Khan, I., Majib, Y., Ullah, R., Rana, O.: Blockchain applications for internet of things—a survey. Internet Things **27**, 101254 (2024). https://doi.org/10.1016/j.iot.2024.101254
3. Li, W., Bu, J., Li, X., Peng, H., Niu, Y., Zhang, Y.: A survey of DeFi security: challenges and opportunities. J. King Saud Univ.-Comput. Inf. Sci. **34**(10), 10378–10404 (2022)
4. Pradhan, N.R., et al.: A novel blockchain-based healthcare system design and performance benchmarking on a multi-hosted testbed. Sensors **22**(9), 3449 (2022)
5. Zou, W., et al.: Smart contract development: challenges and opportunities. IEEE Trans. Softw. Eng. **47**(10), 2084–2106 (2019)
6. Salama, R., Altorgoman, A., Al-Turjman, F.: An overview of how AI, Blockchain, and IoT are making smart healthcare possible. In: Computational Intelligence and Blockchain in Complex Systems, pp. 255–267 (2024)
7. Omidian, H.: Synergizing blockchain and artificial intelligence to enhance healthcare. In: Drug Discovery Today, p. 104111 (2024)
8. Zheng, C., et al.: Decentralized artificial intelligence in construction using blockchain. Autom. Constr. **166**, 105669 (2024)
9. Dagher, G.G., Mohler, J., Milojkovic, M., Marella, P.B.: Ancile: privacy-preserving framework for access control and interoperability of electronic health records using blockchain technology. Sustain. Urban Areas **39**, 283–297 (2018)
10. Zhang, D., Wang, S., Zhang, Y., Zhang, Q., Zhang, Y.: A secure and privacy-preserving medical data sharing via consortium blockchain. Secur. Commun. Netw. **2022**(1), 2759787 (2022)
11. Angraal, S., Krumholz, H.M., Schulz, W.L.: Blockchain technology: applications in health care. Circ. Cardiovasc. Qual. Outcomes **10**(9), 003800 (2017)
12. Alharthi, S., Cerotti, P.R., Far, S.M.: An exploration of the role of blockchain in the sustainability and effectiveness of the pharmaceutical supply chain. J. Supply Chain Customer Relat. Manag. **2020**(2020), 1–29 (2020)
13. Krittanawong, C., et al.: Artificial intelligence-powered blockchains for cardiovascular medicine. Can. J. Cardiol. **38**(2), 185–195 (2022)
14. Jaya, R.M., Rakkhitta, V.D., Sembiring, P., Edbert, I.S., Suhartono, D.: Blockchain applications in drug data records. Procedia Comput. Sci. **216**, 739–748 (2023)
15. Holzinger, A., Haibe-Kains, B., Jurisica, I.: Why imaging data alone is not enough: AI-based integration of imaging, omics, and clinical data. Eur. J. Nucl. Med. Mol. Imag. **46**(13), 2722–2730 (2019)
16. Deng, C., Liu, Y.: A deep learning-based inventory management and demand prediction optimization method for anomaly detection. Wirel. Commun. Mob. Comput. **2021**(1), 9969357 (2021)
17. Chen, H.S., Jarrell, J.T., Carpenter, K.A., Cohen, D.S., Huang, X.: Blockchain in healthcare: a patient-centered model. Biomed. J. Sci. Tech. Res. **20**(3), 15017 (2019)
18. Kantasa-Ard, A., Nouiri, M., Bekrar, A., Cadi, A., Sallez, Y.: Machine learning for demand forecasting in the physical internet: a case study of agricultural products in Thailand. Int. J. Prod. Res. **59**(24), 7491–7515 (2021)

19. Tariq, N., Qamar, A., Asim, M., Khan, F.A.: Blockchain and smart healthcare security: a survey. Procedia Comput. Sci. **175**, 615–620 (2020)
20. Jafri, R., Singh, S.: Blockchain applications for the healthcare sector: uses beyond bitcoin. In: Blockchain Applications for Healthcare Informatics, pp. 71–92. Elsevier (2022)
21. Agbo, C.C., Mahmoud, Q.H., Eklund, J.M.: Blockchain technology in healthcare: a systematic review. Healthcare **7**, 56 (2019)
22. Agbo, C.C., Mahmoud, Q.H.: Blockchain in healthcare: opportunities, challenges, and possible solutions. IJHISI **15**(3), 82–97 (2020) IJHISI **15**(3), 82–97 (2020) IJHISI **15**(3), 82–97 (2020) IJHISI **15**(3), 82–97 (2020) IJHISI **15**(3), 82–97 (2020)
23. Shi, S., He, D., Li, L., Kumar, N., Khan, M.K., Choo, K.-K.R.: Applications of blockchain in ensuring the security and privacy of electronic health record systems: a survey. Comput. Secur. **97**, 101966 (2020)
24. Russell, S.J., Norvig, P.: Artificial Intelligence: A Modern Approach. Pearson (2016)
25. Neapolitan, R.E., Jiang, X.: Artificial Intelligence: With An Introduction to Machine Learning. CRC Press (2018)
26. Gifari, M.W., Samodro, P., Kurniawan, D.W.: Artificial intelligence toward personalized medicine. Pharmaceut. Sci. Res **8**(2), 1 (2021)
27. Ting, D., et al.: Artificial intelligence and deep learning in ophthalmology. Br. J. Ophthalmol. **103**(2), 167–175 (2019)
28. Yousuf, A., Khursheed, B., Rahman, R., Hamadani, H., Hamadani, A.: AI in healthcare: pioneering innovations for a healthier tomorrow. In: A Biologist's Guide to Artificial Intelligence, pp. 67–78. Elsevier (2024)
29. Väänänen, A., Haataja, K., Vehviläinen-Julkunen, K., Toivanen, P.: AI in healthcare: a narrative review. F1000Research **10**, 6 (2021)
30. Kalra, N., Verma, P., Verma, S.: Advancements in AI based healthcare techniques with focus on diagnostic techniques. Comput. Biol. Med. **179**, 108917 (2024)
31. Pujari, C., Muniyal, B., Rao, A., Sadiname, V., Rajarajan, M., et al.: Identity resilience in the digital health ecosystem: a key recovery-enabled framework. Comput. Biol. Med. **167**, 107702 (2023)
32. Ali, S., et al.: Metaverse in healthcare integrated with explainable ai and blockchain: enabling immersiveness, ensuring trust, and providing patient data security. Sensors **23**(2), 565 (2023)
33. Drira, I., et al.: Eye-rubbing detection tool using artificial intelligence on a smartwatch in the management of keratoconus. Transl. Vis. Sci. Technol. **13**(12), 16 (2024)
34. Azaria, A., Ekblaw, A., Vieira, T., Lippman, A.: MedRec: using blockchain for medical data access and permission management. In: 2016 2nd International Conference on Open and Big Data (OBD), pp. 25–30. IEEE (2016)
35. Zhang, P., White, J., Schmidt, D.C., Lenz, G., Rosenbloom, S.T.: FHIRChain: applying blockchain to securely and scalably share clinical data. Comput. Struct. Biotechnol. J. **16**, 267–278 (2018)
36. Zheng, Z., Xie, S., Dai, H.-N., Chen, X., Wang, H.: Blockchain challenges and opportunities: a survey. Int. J. Web Grid Serv. **14**(4), 352–375 (2018)
37. Gunning, D., Aha, D.: Darpa's explainable artificial intelligence (XAI) program. AI Mag. **40**(2), 44–58 (2019)

Analysis of Customer Decisions: To Renew or Not Renew Home Insurance Contracts

Fatma Manlaikhaf[1](✉), Mourad Nachaoui[2], and Abdelghani Ghazdali[1]

[1] EMI, FST, USMS, 23000 Benimellal, Morocco
manlaikhaffatma@gmail.com, a.ghazdali@usms.ma
[2] LIPIM, ENSA, USMS, 25000 Khouribga, Morocco
m.nachaoui@usms.ma

Abstract. Home insurance provides essential financial protection against potential damages to insured properties. While customers initially choose specific coverages when purchasing a policy, insurance companies have noted that many clients face the decision of whether to renew their contracts or let them lapse. During the renewal process, some customers modify their coverage by adding protections, while others may choose not to renew due to factors like premium costs or changing needs. Understanding the factors that drive these renewal decisions is crucial for insurers to improve customer retention strategies. This study aims to analyze the factors influencing customers' decisions to renew or not renew their insurance contracts. By examining the initial coverages chosen at the contract's inception and any modifications made during the renewal process, the research seeks to identify key determinants that drive these renewal decisions.

With the growing role of artificial intelligence (AI), particularly machine learning (ML) and data analysis, this research aims to leverage these technologies to better assess risk factors and predict customer behavior. By employing machine learning models to analyze large datasets, insurers can enhance their ability to encourage clients to opt for more comprehensive coverage from the outset, thereby remaining competitive in a dynamic global market.

The objective of this study was to predict whether customers would cancel or renew their home insurance policies. This decision often arises when a policy expires, typically due to reasons such as non-payment or failure to renew within the contract's timeframe. Using a real-world dataset from home insurance companies, spanning the years 2007 to 2012, key policy attributes, building characteristics, geographical zones, coverage benefits, and risk factors were analyzed.

Various machine learning algorithms, including Decision Tree (DT), K-Nearest Neighbors (KNN), Support Vector Machine (SVM), and XGBoost (XG), were applied to the dataset, which was divided into 70% for training and 30% for testing.

M. Nachaoui, A. Ghazdali—Contributed equally to this work.

A. Laghrib and A. Ghazdali (Eds.): ICMDS 2024, LNNS 1466, pp. 164–173, 2025.
https://doi.org/10.1007/978-3-031-94802-2_16

Keywords: Home insurance · machine learning · Artificial intelligence · ML models

1 Introduction

The insurance industry relies heavily on data to make informed decisions about policy issuance and customer management. As data continuously evolves, insurance companies must adapt their strategies to remain competitive [1,2]. Among various types of insurance, home insurance plays a crucial role in safeguarding residential properties against unexpected incidents, such as fires, floods, and gas explosions [3,4]. Home insurance is not only essential for protecting high-value assets but also provides financial relief for homeowners, particularly those with outstanding mortgages, by covering repair costs or debt payments in the event of a disaster [5,6].

This study focuses on the analysis of home insurance policies, specifically examining the factors that influence policyholders to cancel or renew their policies. Using a comprehensive dataset containing 66 variables related to residential insurance, this research aims to identify patterns in customer behavior and predict renewal decisions. Understanding these patterns allows insurers to tailor their offerings more effectively and enhance customer retention strategies. By analyzing historical data, such as premium payments and policy renewal trends, we can provide insights into the key drivers behind customer decisions at the time of renewal. Understanding termination decisions is crucial for insurance companies, as it enables them to better anticipate customer behaviors and improve retention strategies. By identifying the factors that influence the cancellation of home insurance contracts, insurers can adjust their offerings to meet customers needs and expectations. Furthermore, this in-depth knowledge helps develop pricing policies and tailored incentives to encourage renewals. Therefore, this study aims not only to analyze the determinants of contract termination but also to provide predictive tools that enable companies to enhance their competitiveness in an ever-evolving market [7].

2 Methodology

This study utilized a real-world home insurance dataset sourced from insurance companies. The data was collected between 2007 and 2012, comprising 256,173 rows and a total of 66 variables studied. For this study, R x64 4.0.2 is used for implementing the models. For classification, we used accuracy, precision, recall, specificity, F1-score and AUC as measures of evaluation.

2.1 DATA - Variables Description

Descriptions of some variables studied are presented in Table 1.

The data analysis model utilizes various algorithms, including Decision Tree (DT), K-Nearest Neighbors, Support Vector Machine, and XGBoost. The choice

Table 1. Variables Description.

Variable	Description
Quote_Date	Day where the quotation was made
Cover_Start	Beginning of the cover payment
Claim3Years	3 last years loss
P1empStatus	Client's professional status
Payment_Method	Method of payment
Payement_Frequency	Frequency of payment
Hp3AddonPost_Ren	Option "HP3" included after renewal
Mta_Flag	Mid-Term Adjustment indicator
Mta_Fap	Bonus up to date of Adjustment
Mta_Aprp	Adjustment of the premium for Mid-Term Adjustmen
Mta_Date	Date of Mid-Term Adjustment
LastAnnPrem_Gross	Premium - Total for the previous year
Pol_Status	Police status
Police	Police number

of machine learning models used in this study namely, Decision Tree, Support Vector Machine (SVM) [10–13], and XGBoost-is motivated by their capacity to handle complex and heterogeneous data, as encountered in insurance contract analysis. The Decision Tree model is well-suited to categorical data and provides an intuitive interpretation of decisions made at each level of the tree, facilitating the explanation of renewal factors. SVM, with its various kernels (linear, polynomial, Gaussian), is particularly effective for class separation in high-dimensional spaces, making it relevant for distinguishing customers based on their likelihood to renew. Finally, XGBoost, which is based on a boosting approach, is known for its high performance and its ability to handle imbalanced datasets by reducing generalization error [14–17]. These models thus offer a combination of accuracy, robustness, and interpretability that meets the specific needs of renewal decision analysis in home insurance.

The dataset is split into 70% for training and 30% for testing. We began by examining and understanding the current state of policyholders, both before and after data preparation, the data is categorized into four types: active policies (Live), expired policies (Lapsed), canceled policies (Cancelled), and unknown statuses (Unknown). Initially, there were significant gaps in the data, accounting for 26.2%, as shown in Fig. 1. Upon further investigation, it became clear that some of these gaps prevented further analysis. As a result, those rows were removed (Fig. 2), and missing data in the remaining records were filled using various techniques. Pie charts were generated to visualize two new variables: Resiliated (policy cancellation) and Non_Resiliated (renew the policy). Figure 3

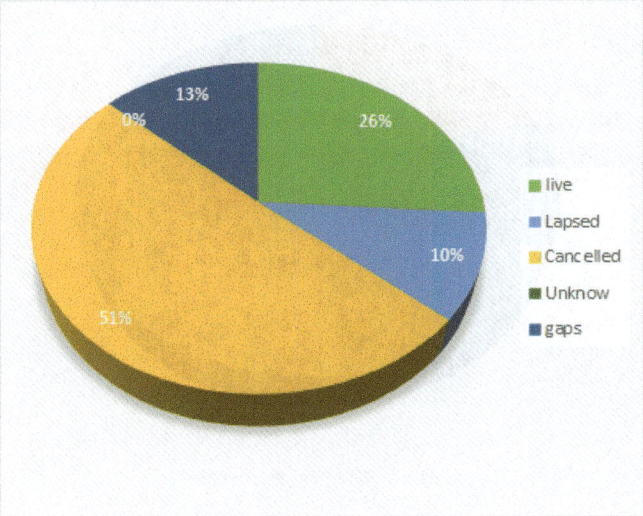

Fig. 1. The four types of data.

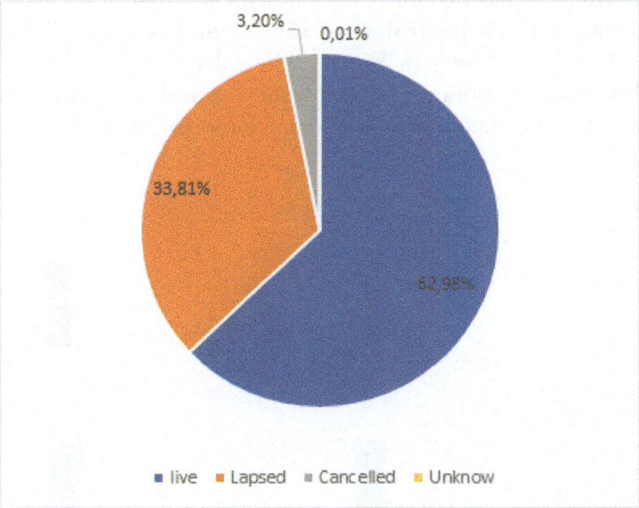

Fig. 2. Data Set After Removing Missing Values.

illustrate the percentages for policy renewal, where '0' indicates that the customer canceled the policy, and '1' indicates that the customer renewed the policy.

The next step involves analyzing the relevant data by examining the relationships between variables using Exploratory Data Analysis (EDA), focusing on Qualitative or Categorical Variables.

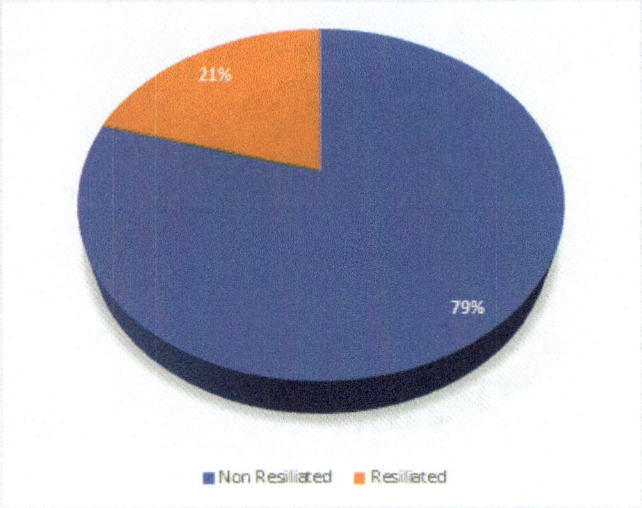

Fig. 3. Distribution of target values.

The data will be classified into two categories: first, the selected variables that can be analyzed, as illustrated in Fig. 4, and second, the variables that cannot be analyzed, as shown in Fig. 5. The selection method is based on the percentage difference in policy contract renewal and non-renewal rates. If the difference exceeds 1.5%, the variable is considered for analysis.

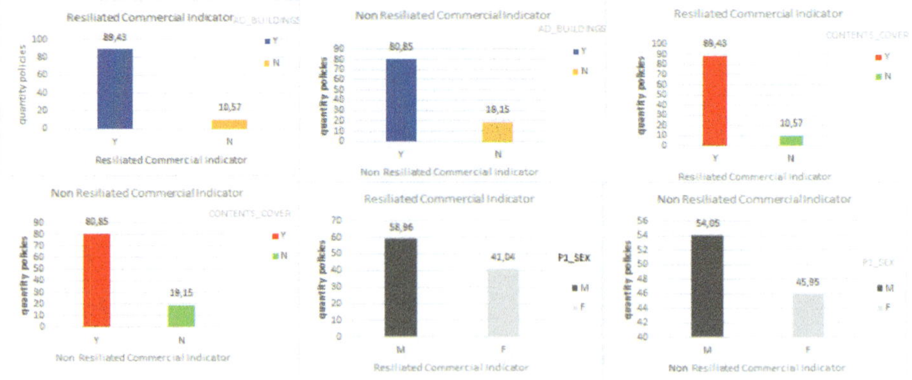

Fig. 4. Overview of Selected Analyzed Variables.

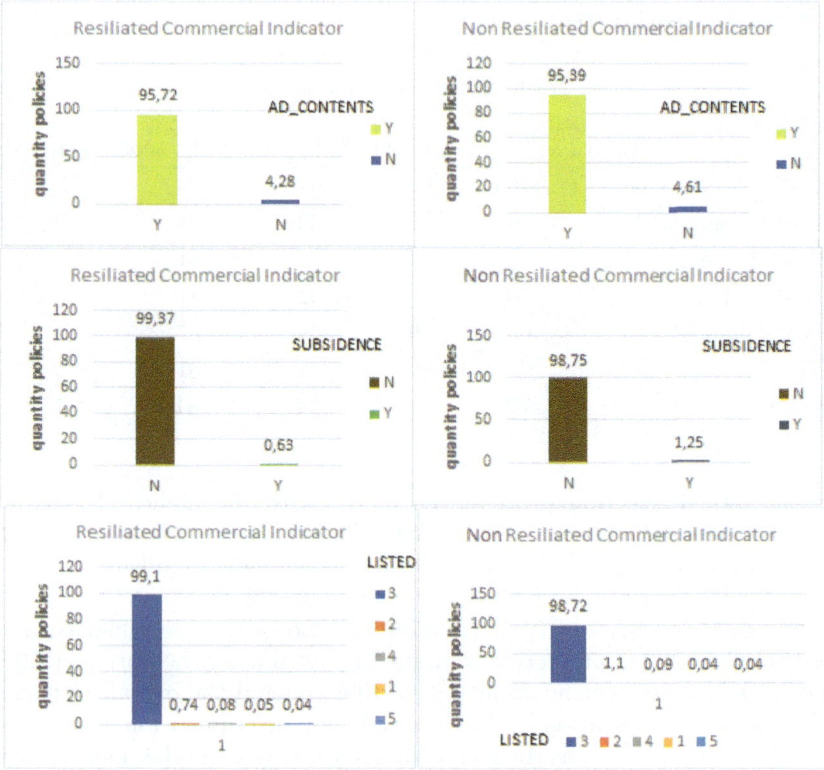

Fig. 5. Selected and non-analyzed variables.

The quantitative variables were analyzed by calculating the mean and median, filling in missing values, and converting certain variables into categories. After selecting the relevant features, 28 variables were identified, including: Claim3Years, P1_Emp_Status, P1_Pt_Emp_Status, Ad_Buildings, Sum_Insured_Buildings, Sum_Insured_Contents, Contents_Cover, Spec_Sum_Insured, Spec_Item_Prem, P1_Mar_Status, P1_Sex, Appr_Locks, Max_Days_Unocc, Sec_Disc_Req, Payment_Method, Legal_Addon_Pre_Ren, Legal_Addon_Post_Ren, Home_Em_Addon_Pre_Ren, Home_Em_Addon_Post_Ren, Garden_Addon_Pre_Ren, Garden_Addon_Post_Ren, Keycare_Addon_Pre_Ren, Keycare_Addon_Post_Ren, Hp2_Addon_Pre_Ren, Hp2_Addon_Post_Ren, Mta_Fap, Mta_Aprp, and Last_Ann_Prem_Gross. These prepared data were then utilized to make predictions.

3 Results and Discussion

Table 2. Model performance.

Model	Accur.(%)	Prec.(%)	Recall(%)	Specifi.(%)	F1-score(%)	RANK
KNN	66	100	61	0	76	6
SVM (Gaussian)	70	70	97	30	81	2
SVM (Linear)	69	69	97	27	81	4
SVM (Poly)	70	70	96	32	81	3
SVM (Sigmoid)	55	65	69	68	67	7
DT	68	68	99	21	81	5
XGB	**72**	**72**	**93**	**34**	**81**	**1**

From the model selection process, the model with the highest F1-Score was XGBoost, followed by Decision Tree, SVM (Gaussian), SVM (Linear), and SVM (Poly), each achieving a certain F1-Score percentage. In terms of accuracy, XGBoost performed the best with 71% accuracy. For precision, XGBoost also led with a precision score of 72%. Therefore, XGBoost was identified as the most suitable model for this prediction, ranking highest among the options (Table 2).

The ROC curves are shown in Fig. 6. ROC offered the overall performance variable of the classification as its threshold for discrimination (Ariana et al. 2006). The AUC is known as the general quality index of the classifiers. The value of 1 for the AUC means a perfect classifier, while 0.5 means a random classifier. Based on the AUC comparison of the classifiers, the XGB score was 0.707, which was the best, followed by SVM (Gaussian) with 0.686, SVM (Poly)(0.656), KNN (0.62), SVM (Sigmoid) (0.606), SVM (Linear) (0.585), and DT (0.575). These findings suggest that XGB, SVM (Gaussian), SVM (Poly), and KNN had satisfactory performance, whereas DT had a poor performance.

3.1 Variable Importance

Variable importance is a technique that tests the contribution of each independent variable to the outcome prediction. To understand which variables most impact our model, we analyzed feature importance from the XGBoost model, which highlights the variables that significantly influence the model's predictions (Fig. 7).

The variables impacting our model include Payment_Method. Once the important variables are identified, we use the Partial Dependence Plot (PDP) to analyze the marginal effect of each feature on the model's output (probability in classification tasks) [8]. This allows us to observe which features have a stronger or weaker influence on the model's predictions and understand how these features correlate with the model's outcomes, as illustrated in Fig. 8.

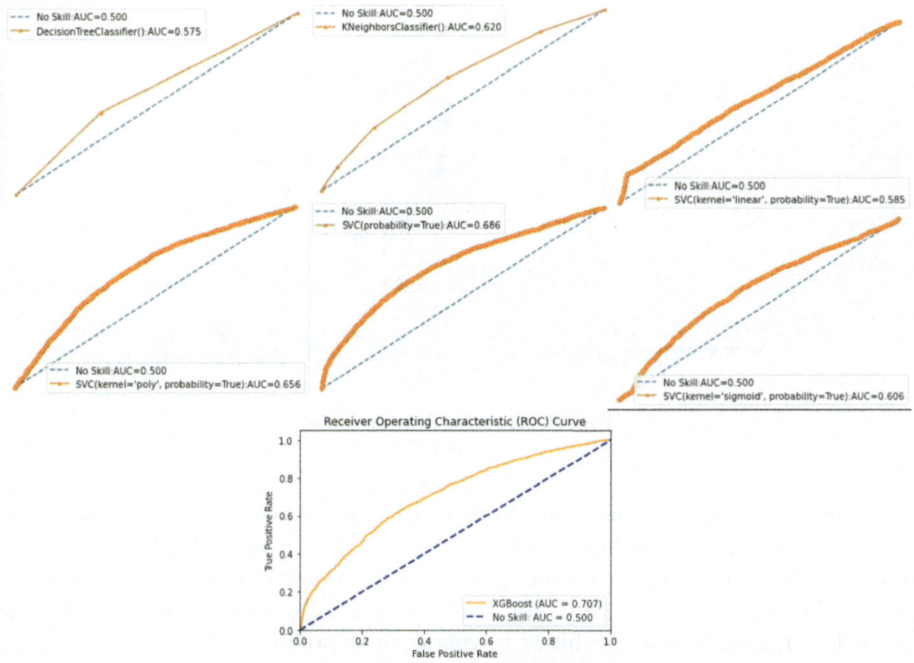

Fig. 6. ROC and calculated AUC obtained for the classifier models of DT, KNN, SVM(Linear), SVM(Poly), SVM(Gaussian), SVM(Sigmoid) et XGB .

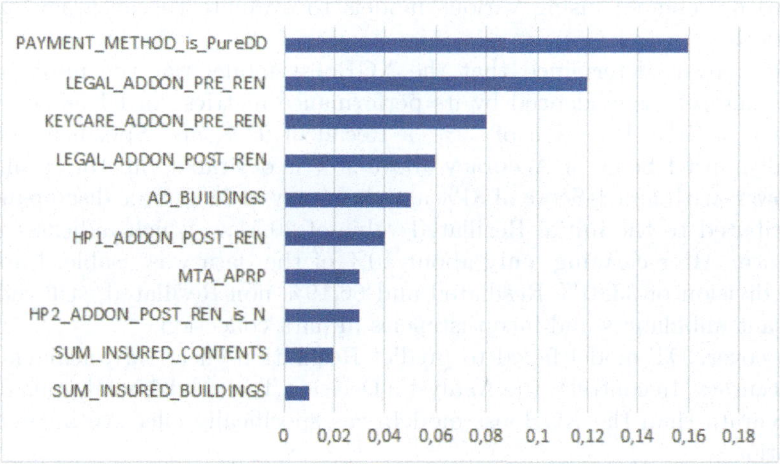

Fig. 7. The rank of features by importance based on the XGBoost model.

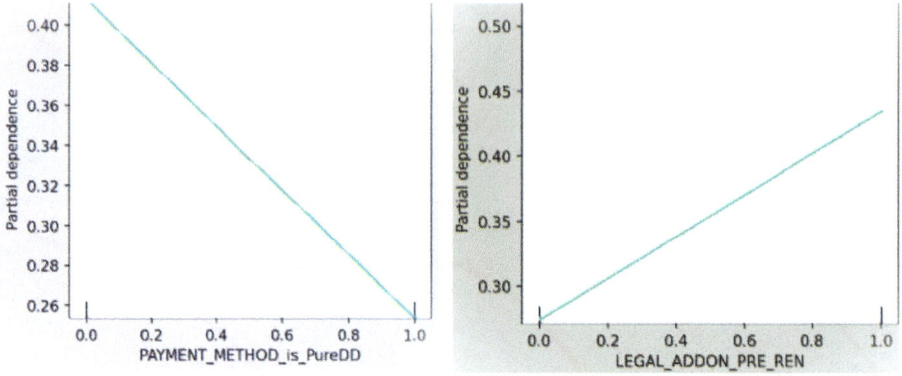

Fig. 8. Variable Importance.

From Fig. 8, it can be observed that when the variable Payment_Method_is_PureDD has a value of 0, the policy is more likely to be resiliated. In other words, if a customer opts not to use the PureDD payment method, the likelihood of policy renewal increases. Additionally, customers who pay legal fees before renewal are more likely to renew their policy.

4 Conclusion

This study focused on predicting whether a home insurance policy would be canceled or renewed, using various models to evaluate the accuracy of these predictions.

The analysis determined that the XGBoost model was the most suitable for this dataset, as evidenced by its performance metrics: an F1-Score of 81%, Accuracy of 71%, Precision of 72%, a Recall of 93%, and Specificity of 34%. Typically, an F1-Score or Accuracy above 90% is desirable, but the results here were lower, with an F1-Score of 81% and Accuracy of 71%. This discrepancy can be attributed to the initial Resiliated value of 20.51%, which indicates a data imbalance. After cleaning, only about 10% of the data was usable, leading to a final division of 33.81% Resiliated and 66.19% non-Resiliated, still reflecting significant imbalances and inconsistencies in data collection.

Moreover, DT model failed to predict Resiliation effectively, achieving only 68% Accuracy. In contrast, the Analyst's Decision Tree model, while only slightly less accurate than the XGBoost model, was specifically effective in predicting Resiliation.

The most impactful variables for the XGBoost model were identified as Payment_Method_is_PureDD and Legal_Addon_Pre_Rent. These findings suggest that customers opting out of the PureDD payment method, or those who pay legal fees before renewal, are more likely to renew their policy contracts.

References

1. Njegomir, V., Ćirić, J.: Risk modeling in the insurance industry. Strateg. Manag. **17**(1), 53–60 (2012)
2. Fitzpatrick, S.: Fear is the key: a behavioral guide to underwriting cycles. Connecticut Insur. Law J. **10**(2), 255–274 (2004)
3. Schlesinger, H.: The theory of insurance demand. In: Handbook of Insurance, pp. 167–184. Springer (2013)
4. Kunreuther, H.: The role of insurance in reducing losses from extreme events: the need for public-private partnerships. Geneva Pap. Risk Insur.-Issues Pract. **40**(4), 741–762 (2015)
5. Bouwer, L.M.: Have disaster losses increased due to anthropogenic climate change? Bull. Am. Meteor. Soc. **92**(1), 39–46 (2011)
6. Michel-Kerjan, E.O., Kunreuther, H.: Redesigning flood insurance. Science **333**(6041), 408–409 (2011)
7. Harrison, T., Ansell, J.: Customer retention in the insurance industry: using survival analysis to predict cross-selling opportunities. J. Fin. Serv. Mark. **6**, 229–239 (2002)
8. Goldstein, A., Kapelner, A., Bleich, J., Pitkin, E.: Peeking inside the black box: visualizing statistical learning with plots of individual conditional expectation. J. Comput. Graph. Stat. **24**(1), 44–65 (2015)
9. Nachaoui, A., Hakim, A., Laghrib, A. (eds.): JANO'13 2021. SPMS, vol. 372. Springer, Cham (2021). https://doi.org/10.1007/978-3-030-83442-5
10. Nachaoui, M., Afraites, L., Hadri, A., Laghrib, A.: A non-convex non-smooth bilevel parameter learning for impulse and gaussian noise mixture removing. Commun. Pure Appl. Anal. **21**(4), 1249–1291 (2022)
11. El Malki, I., El Hakoume, A., Laghrib, A., Hadri, A.: Improved image denoising via self-supervised Weickert operator learning and plug-and-play learned primal dual. Neurocomputing **620**, 129267 (2025)
12. Ghazdali, A., Ourdou, A., Hakim, M., Laghrib, A., Mamouni, N., Metrane, A.: Robust approach for blind separation of noisy mixtures of independent and dependent sources. Appl. Comput. Harmon. Anal. **60**, 426–445 (2022)
13. Laghrib, A., Ghazdali, A., Hakim, A., Raghay, S.: A multi-frame super-resolution using diffusion registration and a nonlocal variational image restoration. Comput. Math. Appl. **72**(9), 2535–2548 (2016)
14. Hakim, M., Ghazdali, A., Laghrib, A.: A multi-frame super-resolution based on new variational data fidelity term. Appl. Math. Model. **87**, 446–467 (2020)
15. Ghazdali, A., Hakim, A., Laghrib, A., Mamouni, N., Raghay, S.: A new method for the extraction of fetal ECG from the dependent abdominal signals using blind source separation and adaptive noise cancellation techniques. Theor. Biol. Med. Model. **12**, 1–20 (2015)
16. Ghazdali, A., Hadri, A., Laghrib, A., Nachaoui, M.: A blind Poisson-Gaussian noise separation using learning copula densities. Cir. Syst. Sig. Process. **42**(11), 6564–6590 (2023)
17. Ghazdali, A., Hadri, A., Laghrib, A., Nachaoui, M.: Poisson noise and Gaussian noise separation through copula theory. Multimedia Tools Appl. **83**(26), 67927–67952 (2024)

Forecasting FTSE CSE: A Neural Network Model Comparison of GRU, 1D-CNN and LSTM Models

Ayman Mazzouz[1(✉)] and Abdelali Elmoufidi[2]

[1] Data4earth Laboratory, Faculty of Sciences and Technics, Sultan Moulay Slimane University, Beni Mellal, Morocco
mazzouzayman@gmail.com

[2] Prospective Research Laboratory in Economics and Management, National School of Commerce and Management, Hassan 2 University, Casablanca, Morocco
a.elmoufidi@usms.ma

Abstract. Financial time series forecasting is a crucial task for financial institutions, investors, and academics. Indeed, financial markets grow increasingly interdependent and highly responsive to both economic and non-economic events. The ability of market participants to accurately manage market spillover determines how well they perform. The goal of implementing machine learning algorithms is to improve the forecasting process's quality. The architecture of the RNNS family is designed to be optimal for time series forecasting. In this study, we compare three popular machine learning algorithms: CNN, LSTM, and GRU to predict the FTSE CSE index returns. The results of our model comparison indicate that the GRU model outperforms the others, providing more realistic and reliable forecasts based on MAE and RMSE metrics. While all models visually follow the original data trends closely, GRU demonstrates a superior performance in terms of accuracy, as evidenced by the error metrics.

Keywords: Stock prediction · LSTM · CNN · GRU · FTSE

1 Introduction

Trading and investing in the stock market offers substantial opportunities and has captivated investors worldwide. This process involves the trading, buying, and selling of stocks, which represent shares of companies. Participants range from institutional to individual retail investors, all aiming to maximize returns and profits. However, accurately predicting stock market movements remains a formidable challenge, especially for high-risk investors who take substantial financial risks with the hope of significant returns [1]. Traditionally, these investors have relied on technical [2] and fundamental analyses [3], which involve studying company financials, economic factors, and statistical trends derived from trading activity. Recent advances in technology have markedly improved

A. Laghrib and A. Ghazdali (Eds.): ICMDS 2024, LNNS 1466, pp. 174–183, 2025.
https://doi.org/10.1007/978-3-031-94802-2_17

predictive capabilities. Machine learning algorithms, for instance, can analyze vast datasets to uncover patterns often hidden from human analysts [4]. By leveraging mathematical computations and sentiment analysis, these algorithms facilitate more precise market trend predictions using quantitative models [3]. Technologies such as Long Short-Term Memory (LSTM) networks, Gated Recurrent Units (GRU), and Convolutional Neural Networks (CNN) have been particularly transformative in financial forecasting, as they enable the identification of intricate patterns within data.

In this study, we explore the potential of these advanced neural network architectures in predicting movements within the Moroccan stock market, using the FTSE CSE index as a benchmark. This research investigates different combinations of RNN-based and CNN models to improve forecasting precision. The remainder of this paper is structured as follows: Sect. 2 provides a brief related work, Sect. 3 outlines the methodology, Sect. 4 presents results and discusses findings, and Sect. 5 draws key conclusions.

2 Related Work

The FTSE CSE Morocco index serves as a key benchmark for companies on the Casablanca Stock Exchange, offering investors insights into Morocco's industry and capital performance by focusing on liquid securities. It includes securities with a free float above 5% or a market capitalization exceeding 1% of the total index for those with 5–15% free float, ensuring representativeness. Updated in real time, it provides reliable data for market analysis and predictive modeling [5].

In financial forecasting, sequential data like stock prices benefit from models designed to handle time dependencies. Recurrent Neural Networks (RNNs) are ideal for such tasks but face challenges with long-term dependencies due to the vanishing gradient issue. LSTM networks developed to overcome this wich use memory cells and gates to retain information over extended sequences [6]. Widely applied in fields like NLP and financial forecasting, LSTMs are particularly effective in capturing non-linear patterns [7,8] and outperform traditional RNNs in predicting stock market trends, enhancing accuracy in financial forecasting [9].

An alternative to LSTM, the GRU introduced in 2014, is an RNN variant similar to LSTM but with only two gates (reset and update), simplifying information flow. Unlike LSTM, it lacks a separate cell state, relying on a hidden state, making it faster due to fewer operations [10]. GRU combines LSTM's Forget and Input Gates into a single Update Gate [11], optimizing the structure for efficient long-sequence processing.

In addition to RNN-based models, CNNs have shown promise in time series forecasting. Originally popular for image, video, and natural language processing, and they are effective in time series forecasting. Their local perception and weight-sharing reduce parameters, improving learning efficiency [12]. CNNs link input to output through mapping functions, trained in two stages: forward propagation to pass input through layers and error calculation to adjust weights and minimize prediction errors [13].

Several studies have employed various techniques using deep learning methods to predict the future movement of the stock market. These studies often compare different technologies to evaluate their performance. For instance, the authors in this article [14] compared 1D-CNN and LSTM to measure performance and found that both technologies produced great results, but LSTM was better for accurate stock market prediction. However, the authors in this article [15] after a comparison study between GRU and LSTM, concluded that GRU performed slightly better than LSTM. They supported their findings with another article that preferred GRU over LSTM due to fewer parameters needed and faster results. On the other hand, other article [16] proposed a hybrid model of CNN-LSTM-GRU and concluded that the hybrid model performed better than using each model alone. Moreover, they compared the three models and found that LSTM performed better than GRU and CNN. Article [17] discusses a method for predicting Bitcoin prices using a hybrid model of a one-dimensional Convolutional Neural Network (1D-CNN) and a Long Short-Term Memory (LSTM) network. Another study [18] provides a framework combining Convolutional Neural Networks (CNN) and Long-Short-Term Memory (LSTM) networks to leverage both historical stock price data and leading indicators like options and futures. Both works used 1D-CNN for feature extraction from time-series data, while LSTM was used for temporal regression to capture sequential dependencies.

Some articles not only focus on neural network algorithms but also use statistical methods like ARIMA and GARCH models. They provide hybrid models that integrate neural networks and statistical methods. For example, this article [6] uses a model that integrates Generalized Autoregressive Conditional Heteroskedasticity (GARCH) models with Long Short-Term Memory (LSTM) networks and introduces a new Volume-Up (VU) strategy for distribution manipulation.

3 Methods and Data

3.1 Methods

In our study, we deployed RNN models (LSTM, GRU) and CNN models specifically to forecast the movements of the FTSE CSE Morocco. The models process daily closing prices of the FTSE, treating them as a time series. To evaluate the LSTM, GRU and CNN predictive capabilities, we segmented the available data into two parts: the initial 80% of the dataset is used for training the models, while the subsequent 20% serves as the test set. This division allows for a comprehensive assessment of the model's performance. Furthermore, to thoroughly assess the model's efficiency and effectiveness. This approach not only tests the model's adaptability to different market conditions but also provides a broader understanding of its predictive power. Additionally, we incorporated metrics such as Mean Absolute Error (MAE) and Root Mean Squared Error

(RMSE) in our training and testing analysis. These metrics are crucial for quantifying the model's accuracy and reliability in predicting the MASI's trends, thereby offering a clear insight into the LSTM, GRU and CNN models overall performance.

3.2 Data

Below is the initial phase of the study, downloading historical stock market data for FTSE CSE Morocco from Investing.com in CSV format as Table 1 presents, which may be basic building that can provide a foundation for Predictive modeling compatible with many data analysis tools. The data is then put into discrete columns to enable concentrated analysis of separate variables, standardized by scaling the values of each column for consistency and comparability, and cleaning from invalid values such as 'NaN' entries that would otherwise skew the results. Finally, the relevant data files are integrated together into one dataset in order to comprehensively analyze the data. To do this, it splits the refined dataset into two subsets: an 80% training set to carry out the training of the models and a 20% testing set for model performance evaluation. Feature selection has to be done to identify key attributes, inputs to the neural network, with date and closing price being some of them. It fits three models: an LSTM model with two LSTM layers and a dense layer, a GRU model with two GRU layers and two dense layers, and finally a CNN model comprising one convolutional layer, a pooling and flattening layer, followed by two dense layers. The models take sequential input; the optimizer used is 'adam', and Mean Squared Error (MSE) is used for calculating the loss, while the performance metric used is the Root Mean Squared Error (RMSE) (Fig. 1).

Table 1. Descriptive statistics of FTSE CSE Morocco indicator

Index	Study period	Trading days
FTSE CSE	02/01/2018–09/04/2024	1571

4 Experiments and Results

4.1 Experiments Settings

The Fig. 2 illustrates the FTSE CSE daily closing prices over 1571 days, represented by the blue line. This dataset, collected from Investing.com, spans from 2018 to 2024, and forms the foundation of our predictive modeling analysis using LSTM, GRU, and 1D-CNN architectures.

Dataset Split as Fig. 3 provides: The data was divided into two sets: an 80% training set, enabling the models to learn complex patterns within the FTSE

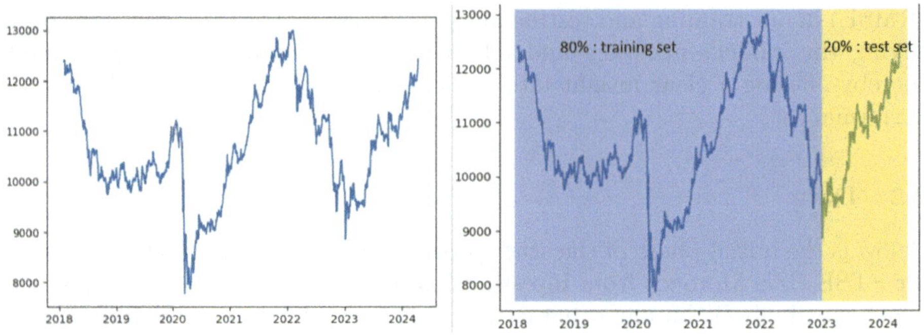

Fig. 1. Time Series of Daily Closing Prices for the FTSE CSE Index

CSE, and a 20% testing set to evaluate each model's predictive power on unseen data.

To assess the predictive accuracy of each model, we used two key evaluation metrics: Mean Absolute Error (MAE) and Root Mean Squared Error (RMSE), as summarized in Table 2. These metrics provide insight into the models' reliability in capturing trends in the FTSE CSE Morocco.

Table 2. Evaluation metrics

Metric	Description	Definition
MAE	Mean Absolute Error	$\text{MAE} = \frac{1}{n}\sum_{i=1}^{n} \lvert y_i - \hat{y}_i \rvert$
RMSE	Root Mean Squared Error	$\text{RMSE} = \sqrt{\frac{\sum_{i=1}^{n}(y_i - \hat{y}_i)^2}{n}}$

Where:

- n is the number of observations.
- yi is the actual value for the ith observation.
- ŷi is the prediction for the ith observation.

To prevent overfitting, we incorporated an early stopping function in our model training, as shown in Listing

Listing 17.1. Early stopping code used in model training to prevent overfitting in Python

```
# Early stopping
early_stopping = EarlyStopping(
    monitor='val_loss', patience=10, min_delta=1e-4,
    mode='min', verbose=1, restore_best_weights=True)
```

The early stopping parameters were set as follows:
Monitor: Tracks validation loss to assess performance improvement.

Patience: Waits for 10 epochs with no improvement before stopping.

Mode: Set to"min," halting training when validation loss stops decreasing.

Restore Best Weights: Restores the model's weights to their best state, optimizing performance.

4.2 Results

4.2.1 Evaluation Metrics To assess the performance of our models in forecasting the FTSE CSE Morocco index, we utilized RMSE and MAE metrics for both training and testing phases. Table 3 below presents the comparative performance metrics of GRU, LSTM, and 1D-CNN models on price prediction.

Table 3. Comparative Performance Metrics of GRU, LSTM and 1D-CNN Models on Price Prediction

	RMSE		MAE	
	Train	Test	Train	Test
GRU	103.534	82.246	68.925	61.064
1D-CNN	105.317	92.695	72.265	68.795
LSTM	98.305	92.735	70.515	68.291

From the table of evaluation metrics that compare the performance of each model we can observe that:

- The GRU model demonstrates strong generalization capabilities, as indicated by a notable improvement in RMSE from training to testing. This suggests that the GRU effectively captures temporal dependencies within the data without overfitting. Furthermore, its superior test MAE further supports its ability to consistently maintain prediction accuracy across unseen data.
- The 1D-CNN model, although effective for extracting spatial features in image and video processing, seems to struggle with capturing sequential dependencies in time series data. The higher RMSE and MAE values in both training and testing phases indicate that this model may be less suitable for this context, possibly due to its focus on localized pattern recognition rather than long-term dependency modeling.
- The LSTM model demonstrates superior learning capabilities during training, achieving the lowest RMSE among the three models. However, its marginally higher test RMSE compared to GRU suggests that while it excels at capturing information over long sequences. Nevertheless, the relatively low-test MAE indicates that the average error remains well-contained, confirming its value in scenarios where minimizing error dispersion is critical.

4.2.2 Comparative Visualization of LSTM, GRU and 1D-CNN Model Predictions From the Fig. 2 provided as result of model prediction, we conclude that:

Due to difficulties with temporal dynamics in time series data, the 1D-CNN model tracks the training data closely but deviates slightly on the test data,

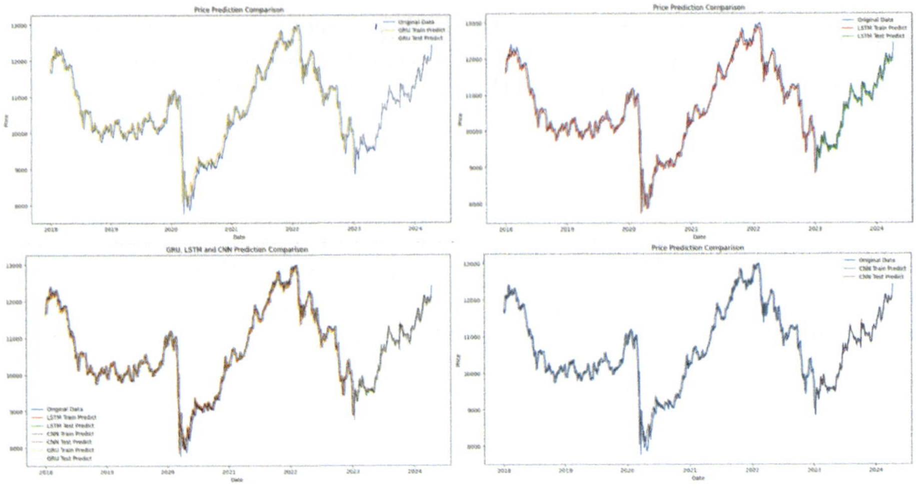

Fig. 2. LSTM, GRU and CNN prediction comparison

suggesting slight overfitting despite early stopping. Effective at identifying long-term dependencies, the LSTM model continuously tracks trends during training and testing. Early stopping helps to avoid overfitting and promotes sound generalization.

A comparison of the three models shows that they all closely match the original data, indicating good generalization and emphasizing the importance of early stopping in preserving prediction accuracy across different architectures.

The Fig. 3 presents the Prediction Error Distributions:

The error distributions of the models are quite clear. Most of the errors of the LSTM model center around zero with a slight skew to the positive, indicating that the predictions are very close to actual values with only a few outliers. The CNN model errors also center around zero but have a larger range and more extreme outliers, which might suggest more variability and particular problems with fitting some of the patterns in the data. The GRU model's errors are more spread out than those of LSTM but less than CNN's, with a slight positive bias to indicate a tendency to overpredict, though outliers are less frequent than in CNN.

The same figure for learning curves provides that:

The learning curves for all the models are all effectively learned with very few cases of overfitting. In the LSTM model, this is evidenced by the steep drop in training loss while the validation loss is stable but with some slight fluctuations throughout, which shows good learning; losses are very close, hence good generalization performance by the model. On the contrary, the CNN model stabilizes a bit faster with hardly any gap between training and validation loss. Whereas the training loss in the GRU model is showing consistent decline and stabilizing, the validation loss also closely follows, hence reflecting that effective learning and generalization are taking place with no significant overfitting.

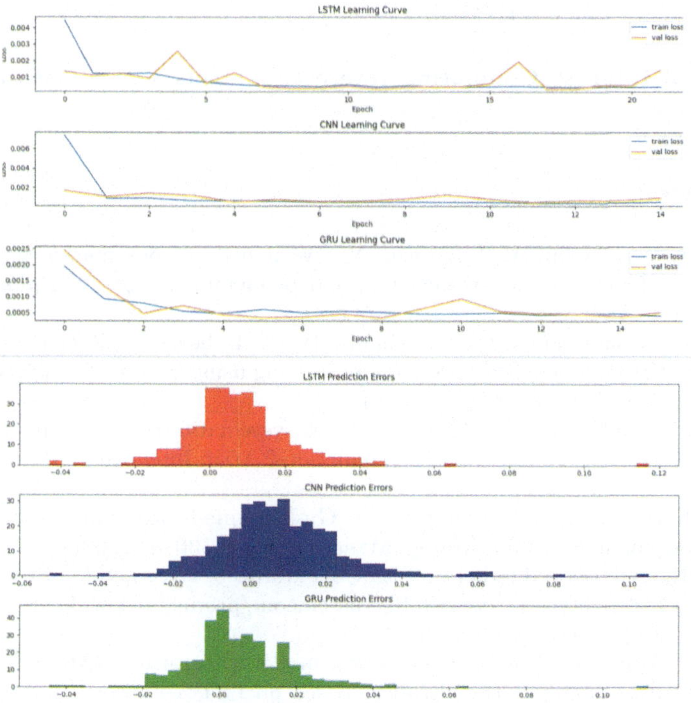

Fig. 3. learning curves and prediction errors

5 Conclusion

This study investigated the performance of three neural network models-GRU, 1D-CNN, and LSTM-in financial time series prediction by providing insight into the strengths and limitations of each model. Thus, the GRU model showed its good generalizing based on lower error rates within the test data as effective performance for capturing sequence dependencies with no overfitting. Spatial data is more fitted to CNN-1D, and therefore, it has less appropriate fit within this context. Time series are sequences with long-term trends, hence, the better suitability of the LSTM model for capturing complex and long-term patterns. It did have the lowest training RMSE, though, but slightly higher test RMSE indicated overfitting, which early stopping helped to counter. In general, the findings have reiterated model selection and training strategically to prevent overfitting, hence improving real-world applications. The study interacts with the design of tailored approaches in the area of financial forecasting and can be considered a base for further research in this domain.

Declarations

Funding. This study was no funding.

Conflicts of Interest/Competing Interests. Authors declare that they have no conflict of interest.

References

1. Kaustia, M., Conlin, A., Luotonen, N.: What drives stock market participation? The role of institutional, traditional, and behavioral factors. J. Bank. Fin. **148**, 106743 (2023)
2. Peng, Y., Albuquerque, P.H.M., Kimura, H., et al.: Feature selection and deep neural networks for stock price direction forecasting using technical analysis indicators. Mach. Learn. Appl. **5**, 100060 (2021)
3. Nti, I.K., Adekoya, A.F., Weyori, B.A.: A systematic review of fundamental and technical analysis of stock market predictions. Artif. Intell. Rev. **53**(4), 3007–3057 (2020)
4. Choudhury, P., Allen, R.T., Endres, M.G.: Machine learning for pattern discovery in management research. Strateg. Manag. J. **42**(1), 30–57 (2021)
5. FTSE Russell: Série d'indices FTSE CSE Morocco, v3.8. London Stock Exchange Group (2022). https://research.ftserussell.com/products/downloads/FTSE_CSE_Morocco_Index_Series_Rules.pdf
6. Koo, E., Kim, G.: A hybrid prediction model integrating GARCH models with a distribution manipulation strategy based on LSTM networks for stock market volatility. IEEE Access **10**, 34743 (2022)
7. Kondiparthy, O., Nimmagadda, S., Venkata, G.K.M., et al.: A concise survey on predicting stock prices. In : 2023 9th International Conference on Advanced Computing and Communication Systems (ICACCS). IEEE, pp. 1182–1186 (2023)
8. De Simone, L., Caputo, E., Cinque, M., et al.: LSTM-based failure prediction for railway rolling stock equipment. Expert Syst. Appl. **222**, 119767 (2023)
9. Gülmez, B.: Stock price prediction with optimized deep LSTM network with artificial rabbits optimization algorithm. Expert Syst. Appl. **227**, 120346 (2023)
10. Prasad, A., Bakhshi, P., Guha, D.: Forecasting the direction of daily changes in the India VIX index using deep learning. IIMB Manag. Rev. **35**(2), 149–163 (2023)
11. GRU enhances the design of gates by merging the Forget Gate and Input Gate of the LSTM into a single Update Gate, streamlining the original cell structure from three gates to two. Additionally, improvements such as cell state fusion have been implemented to optimize the architecture
12. Shah, J., Vaidya, D., Shah, M.: A comprehensive review on multiple hybrid deep learning approaches for stock prediction. Intell. Syst. Appl. **16**, 200111 (2022)
13. Liu, X., Salem, S., Bian, L., et al.: Application of machine learning algorithms in the domain of financial engineering. Alexandria Eng. J. **95**, 94–100 (2024)
14. Rasheed, J., Jamil, A., Hameed, A.A., et al.: Improving stock prediction accuracy using CNN and LSTM. In : 2020 International Conference on Data Analytics for Business and Industry: Way Towards a Sustainable Economy (ICDABI), pp. 1–5. IEEE (2020)

15. Saud, A.S., Shakya, S.: Analysis of look back period for stock price prediction with RNN variants: a case study on banking sector of NEPSE. Procedia Comput. Sci. **167**, 788–798 (2020)
16. Gür, Y.E.: Comparative analysis of deep learning models for silver price prediction: CNN, LSTM, GRU and hybrid approach. Akdeniz İİBF Dergisi **24**(1), 1–13 (2004)
17. Ahmed, O.M., Haji, L.M., Ahmed, A.M., et al.: Bitcoin price prediction using the hybrid convolutional recurrent model architecture. Eng. Technol. Appl. Sci. Res. **13**(5), 11735–11738 (2023)
18. Wu, J.M.T., Li, Z., Herencsar, N., et al.: A graph-based CNN-LSTM stock price prediction algorithm with leading indicators. Multimedia Syst. **29**(3), 1751–1770 (2023)

Optimizing Cryptocurrency Price Prediction with Deep Learning and Evolutionary Tuning

N. Hafidi[1(✉)], Z. Khoudi[1], M. Nachaoui[1], and S. Lyaqini[2]

[1] EMI, FST, Sultan Moulay Slimane University, Beni Mellal, Morocco
{nasreddine.hafidi,zakaria.khoudi,m.nachaoui}@usms.ma
[2] LIPIM, ENSA, Sultan Moulay Slimane University, Khouribgha, Morocco

Abstract. Cryptocurrencies have become a prominent financial asset, but their prices are highly volatile and challenging to predict. Therefore, there is a need for robust models capable of accurately forecasting cryptocurrency prices. ML offers a promising approach for this prediction based on historical price datasets. This study compares several techniques, including Long Short Term Memory and neural networks. Hyperparameters play a crucial role in the performance of these models, and their optimization is essential for achieving high accuracy. To this end, we employ an evolutionary programming approach for hyperparameter optimization. This approach evaluates the performance of these models before and after hyperparameter optimization, aiming to identify the most effective model for cryptocurrency price prediction.

Keywords: cryptocurrency price predictions · time series · deep learning · hyperparameter optimization · genetic algorithm

1 Introduction

Cryptocurrencies have rapidly become vital to the global financial ecosystem [1], offering decentralized, secure, and transparent methods for conducting transactions [2]. However, the high volatility of cryptocurrency prices [3] presents significant challenges for investors and financial analysts [4]. Predicting the future price movements of cryptocurrencies is crucial for making informed investment decisions, as large price swings can lead to substantial gains or losses. Machine Learning (ML) has emerged as a promising approach to tackle this problem by leveraging historical data to train models capable of accurate prediction [5,6], particularly in the domain of financial markets [7]. Numerous ML methods can be used for cryptocurrency price prediction [8]. Each method has its strengths and limitations, and choosing the most suitable algorithm remains challenging. Furthermore, the performance of these algorithms is highly sensitive to hyperparameter tuning [9], which can significantly impact prediction accuracy. Despite

Z. Khoudi, M. Nachaoui and S. Lyaqini—Authors contributed equally to this work.

A. Laghrib and A. Ghazdali (Eds.): ICMDS 2024, LNNS 1466, pp. 184–192, 2025.
https://doi.org/10.1007/978-3-031-94802-2_18

numerous comparative studies in the field [10,11], optimizing hyperparameters remains a critical factor in improving model performance.

This study aims to develop a robust ML architecture for predicting cryptocurrency prices. We will compare various ML techniques, including Recurrent Neural Networks (RNN) [12], Long Short Term Memory (LSTM) [13], Gated Recurrent Unit (GRU) [14] to identify the model that delivers the most accurate price predictions. The performance of each model will be influenced by a set of hyperparameters, which play a crucial role in determining the final results [15]. Even minor hyperparameter changes can significantly influence the model's predictive accuracy. To ensure the model's robustness and optimal performance, we will employ evolutionary programming [16] for hyperparameter optimization [17–20]. This technique will iteratively search for the best set of hyperparameters, ensuring high accuracy. The study will compare the performance of models before and after optimization to evaluate if the optimized models outperform their manually tuned counterparts [21,22].

The primary contribution of this paper is to propose a comprehensive comparison of supervised learning algorithms [23,24], taking into account the influence of hyperparameter settings for each model, to predict future cryptocurrency prices. The selected models will accurately forecast cryptocurrency price trends, enabling more informed investment strategies. This paper also provides a comparative analysis of models with and without hyperparameter optimization, ensuring the selection of the best model for final predictions. The paper is organized as follows: We begin with a problem statement, outlining the mathematical formulation of our model and discussing the influence of hyperparameters on the results. We then present our proposed approach, and conclude with our experimental results.

2 Setting of Problem

Supervised learning [25] is a subfield of ML that relies on labeled data to train algorithms for predicting outcomes and identifying patterns. Hyperparameters are essential settings for a model, influencing its architecture and behavior. Each hyperparameter plays a specific role in shaping the model's performance and predictive power. In the context of our research, the optimization process involves fine-tuning both the model parameters (such as weights and biases) and hyperparameters (including learning rate, batch size, regularization parameters, etc.) of the Neural Network model for cryptocurrency price forecasting. By optimizing these components together, we aim to enhance the model's performance and accuracy in prediction. The general optimization problem can be formally defined as follows:

$$Hyper\_Set^* = \arg\min\{MSE(Y^*, Y_{obs})\}, \quad (1)$$

where:

$$Y^* = argmin(Model(Y_\epsilon)), \quad (2)$$

represents the prediction of the validation dataset, with a given model. Y_{obs} represents the observed data in the validation dataset. $Hyper\_Set^*$ is the optimal set of hyperparameters that gives the minimum of predictions, this set contains the learning rate, size batch, regularization parameter, and number of epochs. MSE: Measures the average of the squares of the errors, defined as follow:

$$MSE = \sum_{i=0}^{n} \frac{(\hat{y}_i - y_i)^2}{n}.$$

By minimizing the equation Eq. 1 we find in the same time, the best set of hyperparameters that gives the best prediction. The best model selection can be based on other evaluation metrics such as the MAE that Measures the average magnitude of errors in a set of predictions without considering their direction. The R^2 Score that Indicates the proportion of variance in the dependent variable that is predictable from the independent variables. And the RMSE that Measures the square root of the average squared differences between predicted and observed values.

To better understand the impact of hyperparameters on model performance, we varied key hyperparameters and observed their effects on our evaluation metrics. In our case, we focused on the window size (W), which represents the length of the historical data used to predict future cryptocurrency prices. Table 1 below demonstrates how varying W affects the performance of the SimpleRNN model, with changes reflected in the values of MAE, R^2 Score, RMSE, and MSE.

Table 1. RNN Model- Changin of metrics value while changing the hyperparameter W

	$Window : W$				
	2	3	5	10	12
MAE	0.0997975	0.08134153	0.08594606	0.09409337	0.098923490
$R^2 Score$	0.2557261	0.58716851	0.48158722	0.55986600	0.307795321
$RMSE$	0.1701316	0.12661289	0.14167843	0.13001940	0.162677944
MSE	0.0289447	0.01603082	0.02007277	0.01690506	0.026464113

3 The Proposed Approach

3.1 Genetic Algorithm

The Genetic Algorithm (GA) [26,27] is a search heuristic inspired by the process of natural selection. It generates high-quality optimization and search problem solutions by applying bio-inspired operators such as mutation, crossover, and selection. For this study, we employed a genetic algorithm to optimize the hyperparameters of our regression model. Below is a simplified pseudo-code of the genetic algorithm used:

Algorithm 1. Genetic Algorithm for Hyperparameter Optimization

```
Input: Population size, mutation rate, crossover rate, termination criteria
Output: Optimal hyperparameter set
Initialize random population of hyperparameter sets
while Termination criteria not met do
    Evaluate fitness of individuals (hyperparameter sets)
    Select fittest individuals for reproduction
    Apply crossover and mutation to generate offspring
    Replace old population with new one
end while
```

The GA begins by initializing a random population of hyperparameter configurations. Each individual in the population represents a possible set of hyperparameters, and their fitness is evaluated using the performance metrics applied to the regression model (MAE and R^2 score). The fittest individuals are selected for reproduction through crossover and mutation operations, generating a new population of hyperparameter sets. This process is repeated until a convergence criterion is met, such as a predefined number of generations or minimal changes in fitness values between generations.

3.2 Methodology

The core idea of our approach is to use a bi-level optimization [28] to simultaneously find the optimal set of hyperparameters for our regression model to ensure robust performance.

We preprocess the dataset, ensuring it is clean, normalized, and split into training and validation sets. The next step involves using the genetic algorithm to search for the best combination of hyperparameters. In our study, the primary hyperparameter variable is the window size W, representing the amount of historical data used for prediction. The genetic algorithm evaluates different values of W by training the model and computing the evaluation metrics for each configuration. Once the algorithm converges on an optimal solution, we use the best hyperparameter set to train the final model. This model's performance is then evaluated on the test dataset, using the same set of metrics to ensure consistency between training, validation, and final evaluation.

4 Results and Discussion

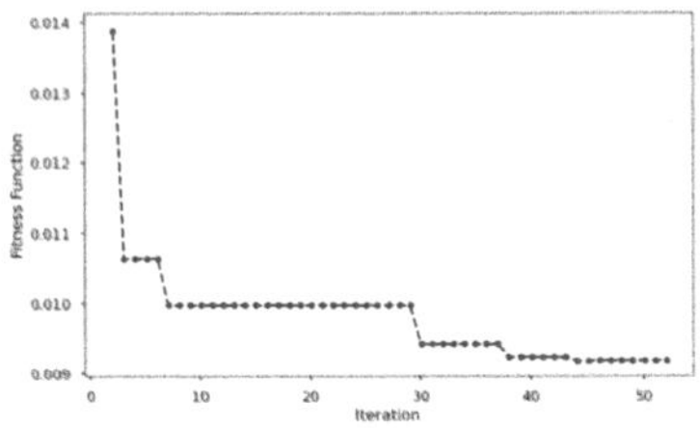

(a) Genetic Algorithm Convergence with LSTM

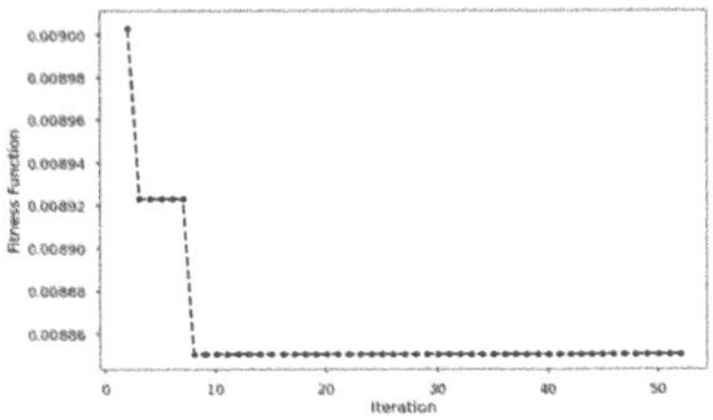

(b) Genetic Algorithm Convergence with Simple RNN

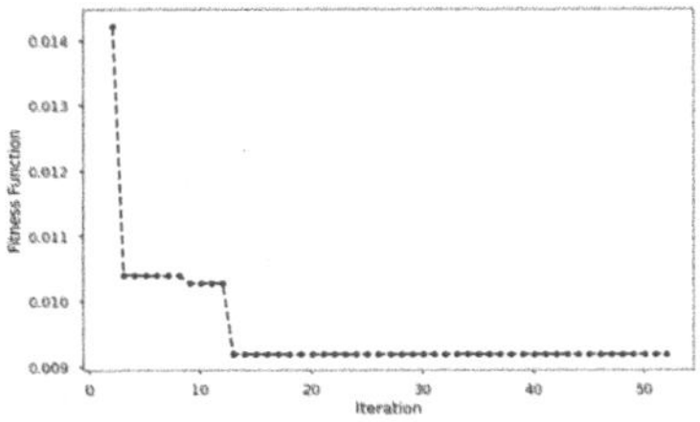

(c) Genetic Algorithm Convergence with GRU

Fig. 1. Convergence of Genetic Algorithm for Different Models.

In Fig. 1a, 1b, and 1c, we observe that the convergence rates of the models vary, with some models reaching optimal hyperparameters faster than others.

This difference in convergence speed can be explained by the complexity of the models and the structure of their hyperparameter spaces. For instance, the hyperparameter optimization process for SimpleRNN is generally faster than LSTM and GRU due to the fewer parameters involved. In contrast, LSTM and GRU, being more complex with multiple gates and internal states, have larger hyperparameter spaces that require more iterations to explore. Moreover, the genetic algorithm's parameters, such as population size and mutation rate, influence how quickly the search converges. These parameters are dynamically adjusted based on the size and complexity of the hyperparameter space of each model. For example, larger populations may be used when optimizing models like LSTM or GRU to ensure better search space coverage. We now pass on to vary the GA parameters for all models and show the convergence of the used models.

Figure 2 shows the convergence graph of the three used models: SimpleRNN, LSTM, and GRU. As we can see in the table below, the convergence process is not the same for all models. For example, while GRU converged on iteration 10, SimpleRNN took more iterations to reach convergence.

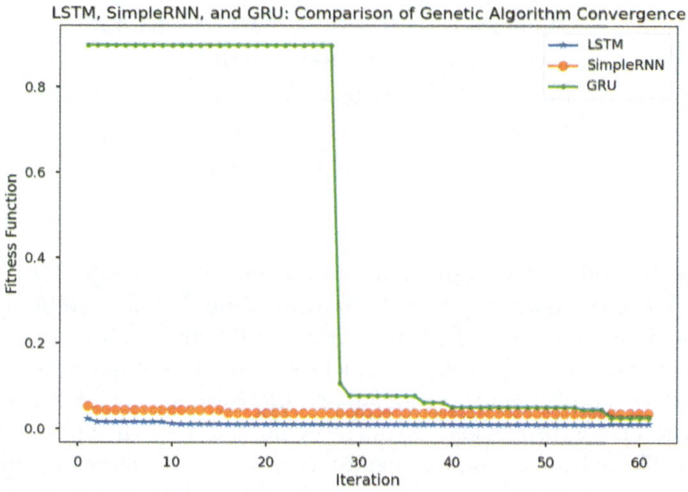

Fig. 2. Genetic Algorithm Convergence for all models

Once the algorithm converges to the optimal hyperparameters, we use them to train the final version of the model. Table 2 compares the model performance before and after the hyperparameter optimization. The results demonstrate significant performance gains in metrics like Root Mean Square Error (RMSE) and accuracy after the genetic algorithm is applied, confirming the effectiveness of the optimization process for SimpleRNN, LSTM, and GRU models.

The Table 2 above presents a detailed comparison of the performance of three models-SimpleRNN, LSTM, and GRU before and after hyperparameter optimization. Each model was evaluated based on three key metrics: RMSE, MSE and MAE. These metrics provide a comprehensive assessment of the models' predictive accuracy and error distribution. In addition to individual metrics, a combined score was calculated to offer a single, weighted measure of performance, incorporating RMSE, MSE, and MAE. The comparison demonstrates the impact of hyperparameter tuning, as reflected in the changes observed in the combined score and individual error metrics.

Table 2. Comparison of SimpleRNN, LSTM, and GRU models before and after hyperparameter optimization.

Model	Metric	Before	After	Improvement (%)
SimpleRNN	MAE	0.0530600	0.00486	-
	MSE	0.8599094	0.000106	-
	RMSE	0.0736495	0.01028	-
	Combined Score	0.22472	0.00662	**97.05**
LSTM	MAE	0.05052158	0.005671	-
	MSE	0.00427882	0.0001373	-
	RMSE	0.0654127	0.0117189	-
	Combined Score	0.04872	0.00759	**84.42**
GRU	MAE	0.116466	0.00705	-
	MSE	0.038736	0.000175	-
	RMSE	0.196815	0.01325	-
	Combined Score	0.14109	0.00878	**93.78**

The results indicate a significant improvement in the performance of all three models following hyperparameter optimization. For the SimpleRNN model, the combined score decreased from 0.22472 to 0.00662, representing a 97.05% improvement. Similarly, the LSTM model saw a reduction in the combined score from 0.04872 to 0.00759, marking an 84.42% improvement, while the GRU model achieved a 93.78% improvement, with the combined score dropping from 0.14109 to 0.00878. These improvements suggest that hyperparameter tuning substantially reduces prediction errors across all models, with the SimpleRNN model showing the most notable enhancement in performance.

The significant improvements in performance across all models highlight the critical role of hyperparameter tuning in enhancing the accuracy and robustness of ML models. By carefully adjusting parameters such as learning rate, window size, and the number of epochs, the models were able to achieve far lower error rates, as indicated by the reductions in MAE, MSE, and RMSE. These results demonstrate that hyperparameter optimization is not only beneficial but essential for achieving optimal performance, especially when dealing with complex time series forecasting tasks. The substantial improvements in the combined scores further emphasize that the effectiveness of the models can be significantly enhanced through proper tuning, leading to more reliable and accurate predictions.

5 Conclusion

This paper explored the application of ML techniques to predict cryptocurrency prices, focusing on the role of hyperparameters in optimizing model performance. We compared RNN, LSTM, and GRU, assessing their predictive capabilities based on evaluation metrics such as MAE, RMSE, and MSE. To address the challenge of manual hyperparameter tuning, we employed a GA for hyperparameter optimization. By automating the search process, the GA efficiently explored the hyperparameter space, identifying the optimal configurations for each model. The results demonstrated that the performance of these models is highly sensitive to hyperparameter choices, and optimizing these parameters with the GA, significantly enhances prediction accuracy. Despite the promising results, our study has several limitations. First, the models were trained on a single dataset, and the generalizability of the results to other cryptocurrency datasets remains untested. Additionally, the computational cost of training and hyperparameter optimization was substantial, especially for LSTM and GRU models, which may pose practical challenges for real-time prediction systems. One potential avenue is to explore more advanced optimization techniques, such as hybrid genetic algorithms combined with other metaheuristics, to further enhance performance and reduce computational costs.

References

1. Almeida, J., Gonçalves, T.C.: A systematic literature review of investor behavior in the cryptocurrency markets. J. Behav. Exp. Fin. **37**, 100785 (2023)
2. Wu, J., Liu, J., Zhao, Y., Zheng, Z.: Analysis of cryptocurrency transactions from a network perspective: an overview. J. Netw. Comput. Appl. **190**, 103139 (2021)
3. Poudel, S., et al.: Cryptocurrency price and volatility predictions with machine learning. J. Mark. Anal. **11**(4), 642–660 (2023)
4. Yin, L., Nie, J., Han, L.: Understanding cryptocurrency volatility: the role of oil market shocks. Int. Rev. Econ. Fin. **72**, 233–253 (2021)
5. Khoudi, Z., Nachaoui, M., Lyaqini, S.: Finding the contextual impacts on students' mathematical performance using a machine learning-based approach. Infocommun. J. Publ. Sci. Assoc. Infocommun. (HTE) **16**(SI), 12–21 (2024)
6. El Malki, I., El Hakoume, A., Laghrib, A., Hadri, A.: Improved image denoising via self-supervised Weickert operator learning and plug-and-play learned primal dual. Neurocomputing **620**, 129267 (2025)
7. Tang, Y., et al.: A survey on machine learning models for financial time series forecasting. Neurocomputing **512**, 363–380 (2022)
8. Khedr, A.M., Arif, I., El-Bannany, M., Alhashmi, S.M., Sreedharan, M.: Cryptocurrency price prediction using traditional statistical and machine-learning techniques: a survey. Intell. Syst. Account. Fin. Manag. **28**(1), 3–34 (2021)
9. Weerts, H.J., Mueller, A.C., Vanschoren, J.: Importance of tuning hyperparameters of machine learning algorithms. arXiv preprint arXiv:2007.07588 (2020)
10. Samson, T.K.: Comparative analysis of machine learning algorithms for daily cryptocurrency price prediction. Inf. Dyn. Appl **3**(1), 64–76 (2024)
11. Hafidi, N., Nachaoui, M., Daoui, C.: A new efficient parallel hierarchical value iteration algorithm using dynamic processor distribution. Int. J. Parallel Emergent Distrib. Syst., 1–17 (2024)
12. Medsker, L.R., Jain, L., et al.: Recurrent neural networks. Design Appl. **5**(2), 64–67 (2001)
13. Sunny, M.A.I., Maswood, M.M.S., Alharbi, A.G.: Deep learning-based stock price prediction using LSTM and bi-directional LSTM model. In: 2020 2nd Novel Intelligent and Leading Emerging Sciences Conference (NILES), pp. 87–92. IEEE (2020)

14. Xiuyun, G., Ying, W., Yang, G., Chengzhi, S., Wen, X., Yimiao, Y.: Short-term load forecasting model of GRU network based on deep learning framework. In: 2018 2nd IEEE Conference on Energy Internet and Energy System Integration (EI2), pp. 1–4. IEEE (2018)
15. Ghazdali, A., Ourdou, A., Hakim, M., Laghrib, A., Mamouni, N., Metrane, A.: Robust approach for blind separation of noisy mixtures of independent and dependent sources. Appl. Comput. Harmon. Anal. **60**, 426–445 (2022)
16. Gul, F., Rahiman, W., Alhady, S., Ali, A., Mir, I., Jalil, A.: Meta-heuristic approach for solving multi-objective path planning for autonomous guided robot using PSO-GWO optimization algorithm with evolutionary programming. J. Ambient. Intell. Humaniz. Comput. **12**, 7873–7890 (2021)
17. Yang, L., Shami, A.: On hyperparameter optimization of machine learning algorithms: theory and practice. Neurocomputing **415**, 295–316 (2020)
18. Hafidi, N., Khoudi, Z., Nachaoui, M., Lyaqini, S.: Enhanced SMS spam classification using machine learning with optimized hyperparameters. Indonesian J. Electr. Eng. Comput. Sci. **37**(1), 356–364 (2025)
19. Lyaqini, S., Nachaoui, M.: Diabetes prediction using an improved machine learning approach. Math. Model. Comput. **8**(4), 726–735 (2021)
20. Hakim, M., Ghazdali, A., Laghrib, A.: A multi-frame super-resolution based on new variational data fidelity term. Appl. Math. Model. **87**, 446–467 (2020)
21. Ourdou, A., Ghazdali, A., Laghrib, A., Metrane, A.: Blind separation of instantaneous mixtures of independent/dependent sources. Cir. Syst. Sig. Process. **40**(9), 4428–4451 (2021)
22. Afraites, L., Hadri, A., Laghrib, A.: A denoising model adapted for impulse and Gaussian noises using a constrained-PDE. Inverse Prob. **36**(2), 025006 (2020)
23. Khoudi, Z., Nachaoui, M., Lyaqini, S.: Identifying the contextual factors related to the reading performance of Moroccan fourth-grade students from a machine learning-based approach. Educ. Inf. Technol. **29**(3), 3047–3073 (2024)
24. Lyaqini, S., Hadri, A., Ellahyani, A., Nachaoui, M.: Primal dual algorithm for solving the Nonsmooth twin SVM. Eng. Appl. Artif. Intell. **128**, 107567 (2024)
25. Lyaqini, S., Nachaoui, M.: Heart failure prediction using supervised machine learning algorithms. In: The International Conference on New Trends in Applied Mathematics, pp. 37–49 (2022). Springer
26. Katoch, S., Chauhan, S.S., Kumar, V.: A review on genetic algorithm: past, present, and future. Multimedia Tools Appl. **80**, 8091–8126 (2021)
27. Nachaoui, M., Afraites, L., Laghrib, A.: A regularization by denoising super-resolution method based on genetic algorithms. Sig. Process. Image Commun. **99**, 116505 (2021)
28. Laghrib, A., Bella, F., Nachaoui, M., Jauberteau, F.: Bilevel optimal parameter learning for a high-order nonlocal multiframe super-resolution problem. Inverse Prob. **40**(1), 015004 (2023)

Unraveling Differences and Similarities in Sustainable Investment Through Case Studies in China, Germany, and the United States: A Multi-algorithm Clustering Network Approach

Mohamed Amine Ouhinou[1(✉)], Abdellah Saoualih[2], Soufiane Lyaqini[3], Salah Eddine Kartobi[1], Stoyan Prodanov[4], and Larbi Safaa[2]

[1] Department of Economic Sciences, Research Laboratory in Energy Economics, Environment and Resources, University of Cadi Ayyad, Marrakesh, Morocco
mohamedamine.ouhinou@ced.uca.ma, se.kartobi@uca.ac.ma
[2] Laboratory for Studies on Resources, Mobility and Attractiveness, Faculty of Letters and Human Sciences, University of Cadi Ayyad, Amerchich - Daoudiate B.P. 2410, 40000 Marrakesh, Morocco
abdellah.saoualih@ced.uca.ma, safaa.larbi@uca.ma
[3] ENSA, Sultan Moulay Slimane University Khouringa, Beni Mellal, Morocco
[4] D. A. Tsenov Academy of Economics, Em. Chakarov 2 Str, 5250 Svishtov, Bulgaria
s.prodanov@uni-svishtov.bg

Abstract. This study offers a comparative analysis of sustainable investment practices in three major economies: China, Germany, and the United States. Drawing on empirical data from the Scopus database (2019–2024), it leverages advanced natural language processing (NLP) techniques combined with a multi-algorithm clustering approach to identify and analyze thematic clusters. The results highlight a global convergence toward the adoption of Environmental, Social, and Governance (ESG) criteria and the integration of Sustainable Development Goals (SDGs) into national policies. However, significant contextual divergences persist: the United States faces structural challenges in adopting green technologies, Germany experienced mixed performance of ESG indices during the COVID-19 crisis, and China struggles with inconsistencies in ESG ratings, threatening financial stability. The originality of this research lies in its dual contribution: methodologically, by integrating NLP and co-word analysis to provide a robust and innovative analytical framework; and empirically, by offering practical recommendations for policymakers and investors to harmonize ESG standards and develop sustainable strategies tailored to the economic and cultural specificities of each context. These findings lay the groundwork for aligning sustainable investment practices with global and local development objectives

A. Saoualih, S. Lyaqini, S.E. Kartobi, S. Prodanov and L. Safaa—Authors contributed equally to this work.

A. Laghrib and A. Ghazdali (Eds.): ICMDS 2024, LNNS 1466, pp. 193–202, 2025.
https://doi.org/10.1007/978-3-031-94802-2_19

while enhancing the coherence of sustainable finance policies on a global scale.

Keywords: Sustainable investment · China · Germany · United States · empirical studies · N-grams · co-word analysis · multi-algorithm clustering network · natural language processing · Scopus

1 Introduction

One of the most effective approaches to addressing the challenges of climate change, social issues, and ethnic conflicts is through the implementation of financial market regulations and the adoption of ethical standards by financial institutions. In the era of globalization, the active involvement of financial markets, intermediaries, and regulators is crucial for ensuring an effective response to these challenges. Sustainable investment integrates financial goals with ESG concerns, aiming to achieve economic returns while conserving natural resources and enhancing social well-being [1,2]. This approach marks a shift towards an economy that places environmental and social challenges at the core of corporate strategies and investment choices [3]. It emerges from an increasing recognition of the traditional model's shortcomings, which typically prioritized profit maximization without regard for environmental and social impacts [4].

The evolution of sustainable investment is shaped by distinct national contexts. In China, early 2000s policies aimed at green growth and initiatives like the "Belt and Road Initiative" sought to address environmental challenges and reduce fossil fuel reliance, though rapid economic growth occasionally conflicted with these goals. In the U.S., sustainable investment has progressed with the financial markets, experiencing fluctuations with changing administrations, and seeing renewed focus under the Biden administration with initiatives like "Build Back Better" and rejoining the Paris Agreement. Germany has long embraced sustainability, initiating policies in the 1990s like "Energiewende" to decrease fossil fuel dependency and promote renewable energy, bolstered by strong European ESG regulations and societal consensus.

Building on the above, this study adopts a comparative framework to analyze thematic trends in sustainable investment research across three major economies: China, Germany, and the United States, during the period 2019–2024. This time frame was chosen to reflect recent global economic shifts and the heightened focus on sustainability following the COVID-19 pandemic. Methodologically, the study integrates advanced natural language processing (NLP) with a multi-algorithmic co-word analysis approach to identify and map thematic clusters. This innovative framework not only enables the detection of patterns in scientific production but also facilitates a deeper understanding of the socio-economic and cultural factors shaping sustainable investment strategies in distinct contexts. By examining the convergence towards ESG criteria and uncovering local divergences, this study provides actionable insights for policymakers and practitioners seeking to harmonize sustainable finance practices at a global level.

This paper is structured as follows: Sect. 2 describes the methodological approach and the techniques used for data processing. Section 3 presents the results on thematic clusters related to sustainable investments in the three countries. In Sect. 4, we discuss the main points of convergence and divergence. Finally, Sect. 5 summarizes the conclusions.

2 Methods and Materials

The first subsection explains how to implement the methodology, while the second subsection describes the data pre-processing steps.

2.1 Co-word Analysis Based on a Multi-algorithm Clustering Network Approach

Co-word analysis allows for the exploration of thematic relationships by examining co-occurrence networks, thereby revealing conceptual groupings. In our study, the multi-algorithmic approach is adopted primarily to select the algorithm offering optimal modularity, ensuring a relevant deduction of interrelations between topics, the identification of specific sub-themes, and the precise tracking of the evolution of empirical research on sustainable investment within the contexts of China, Germany, and the United States.

From a technical standpoint, the first step is to convert the authors' keywords into a list of N-grams, which are contiguous sequences of n elements within a textual expression [5–7]. This process can be mathematically represented by the following formula:

$$N - gram(w_i) = (w_i, w_i + 1, ..., w_i + N - 1) \tag{1}$$

An N-gram refers to a sequence of N elements, where N represents the length of the sequence. In the context of textual expressions, w_i denotes the i-th word, while $w_i + 1, w_i + 2, ..., w_i + N - 1$ refers to the subsequent words, collectively forming a sequence of N consecutive words.

The second step involves using the N-grams generated from the dataset to construct a network where each node represents a unique N-gram. Connections, or edges, between nodes are established when two N-grams appear together in the same keyword set of an article, indicating their co-occurrence. The strength of each connection increases with repeated co-occurrences, thereby reflecting the intensity of the association between the N-grams.

The third step involves applying different algorithms to each bibliographic corpus. Several algorithms were tested, as shown in Table 1, which presents a comparison of modularity values for each country studied (China, Germany, United States). Modularity measures the quality of network division into clusters by quantifying the density of links within clusters compared to those between different clusters. In the case of China, the Leiden algorithm was selected, yielding a modularity value of 0.628, which resulted in the clustering of seven thematic clusters, demonstrating a strong ability to identify coherent topics. For Germany, the Leading Eigenvector and Fast Greedy algorithms showed the best performance, both achieving a modularity value of 0.724, reflecting excellent quality in dividing topics within the corpus, also resulting in seven thematic clusters. Finally, in the United States, the Louvain algorithm was chosen, producing a modularity value of 0.645, which led to the clustering of five thematic clusters, indicating a satisfactory distribution of topics in this context.

Table 1. Comparison of modularity values across clustering algorithms

Algorithms	China	Germany	USA
Leiden [8]	**0.628**	0.715	0.634
Louvain [9]	0.619	0.714	**0.645**
Girvan-Newman [10]	0.412	0.688	0.567
Leading Eigenvector [11]	0.602	**0.724**	0.644
Fast Greedy [12]	0.613	**0.724**	0.642
Label Propagation [13]	0.589	0.709	0.629
Walktrap [14]	0.606	0.710	0.622
Infomap [15]	0.605	0.708	0.642
Spinglass [16]	0.549	0.557	0.603

2.2 Data Handling

This research focused on empirical articles in the fields of economics and management concerning sustainable investment, published between 2019 and 2024 and sourced from the Scopus database. These articles, all in English and at the definitive publication stage, were meticulously selected for their relevance to the national context of each studied country. In China, from an initial selection of 51 articles, 47 were retained for analysis. In Germany, all 28 identified articles were fully included. In the United States, 23 articles were deemed relevant and included in the study.

After collecting the data, we applied a preprocessing procedure to the target data (author keywords) using several NLP techniques. First, word lemmatization was employed to normalize different forms of a word, ensuring that grammatical variations were eliminated and only the core meanings were analyzed for greater accuracy. We then converted all keywords to lowercase to standardize them and prevent duplication due to differences in capitalization. Next, we generated N-grams, which are combinations of consecutive words, to better capture key concepts. Then, we selected the 50 most frequent N-grams to focus on the most commonly used concepts, based on an optimal modularity value.

After pre-processing data, the vectorization stage started. The "Authors' keywords" were converted into numerical vectors using the CountVectorizer tool in this step.

First, a vocabulary V is constructed from all the unique words found in the "Authors' keywords" of the corpus, with N representing the total number of these unique words.

$$V = \{v_1, v_2, \ldots, v_N\} \tag{2}$$

The N -size vocabulary is built from all the unique words in the "Authors' keywords" of the corpus.

Then, for each set of keywords K_i in the corpus, a vector $\mathrm{v_i}$ is generated. This vector has a dimension equal to the total number of unique words in the vocabulary N. Each element $v_{i,j}$ of vector $\mathrm{v_i}$ corresponds to the number of occurrences of word j in the keywords set K_i.

$$K = \{K_1, K_2, \ldots, K_M\} \tag{3}$$

Let K be the set of M keywords sets in the corpus.

Finally, the document-term matrix M is created. This matrix is of size $N \times M$, where N is the total number of unique words in the vocabulary and M is the total number of documents (keywords sets) in the corpus. Each row of the matrix represents a unique word, and each column represents keywords set. The value $m_{i,j}$ in the matrix indicates the number of occurrences of word i in the keywords set j.

$$V_i = [v_{i,1}, v_{i,2}, \ldots, v_{i,N}] \tag{4}$$

Let M be the document-term matrix where, $m_{i,j}$ represents the number of occurrences of word i in the keywords set j.

3 Results

The findings of our methodological approach are shown in Fig. 1, which presents a bidimensional representation of the co-word networks of keywords related to sustainable investment for three countries: China, Germany, and the United States. In the case of China, the Leiden algorithm was selected, yielding a modularity value of 0.628, which resulted in the clustering of seven thematic clus-

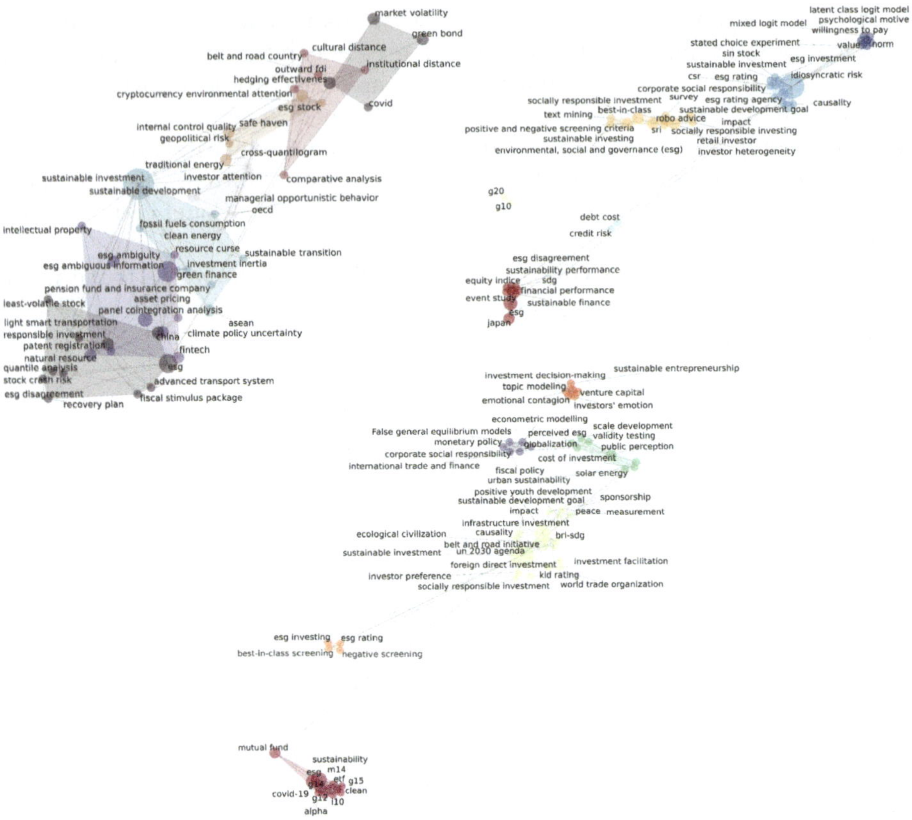

Fig. 1. Bidimensional display of co-word networks of keywords in sustainable investment. Note: The network at the top left illustrates the case of China, the one at the top right represents the case of Germany, while the network at the bottom center corresponds to the case of the United States.

ters. For Germany, the Leading Eigenvector and Fast Greedy algorithms demonstrated the best performance, both achieving a modularity value of 0.724, which resulted in the clustering of seven thematic clusters. In the case of the United States, the Louvain algorithm was utilized, producing a modularity value of 0.645, which led to the formation of five thematic clusters.

Table 2 provides a mapping of the themes for each cluster, created by identifying the associated empirical articles based on the authors' keywords, followed by an analysis of their abstracts. This information was then used to define the main themes for each cluster

Table 2. Mapping thematic trends on sustainable investment

Chinese context	German context	United States context
The impact of ESG factors and the complex interactions between climate policies, financial risks, and sustainable investments	Challenges and opportunities in sustainable investments: Analysis of financial performance and investor behavior	Integrating sustainable investments with global initiatives for a responsible transformation of financial policies and practices
The interactions between sustainable investments, green finance, and their impact on various economic and ecological sectors	Venture capital supports the financing of sustainable startups by shaping investor perceptions and investment decisions	The impact of sustainable investments and climate policies on financial performance and risk management
The impact of ESG factors on financial stability, sustainable investments, and risk management in financial markets and businesses	Investor behavior and sustainability criteria in the analysis of preferences and sustainable investment strategies	Sustainable investments and SDGs in sport for development, green technology, and global initiatives
The impact of investor attention on management practices in the energy sector and the effectiveness of carbon capture and storage technologies to reduce CO2 emissions in the energy industries	The effectiveness of the regulatory framework for sustainable finance in Europe: Transparency, complexity, and mobilization of private capital	Public perception of ESG initiatives and the adoption of green technologies in light of challenges and barriers
Analysis of the reactivity of ESG shocks, their relationship to traditional asset classes, and their impact on financial stability	ESG risk assessment and its impact on financial stability and the cost of capital	Exploring Basel III reforms as catalysts for sustainable investment and the transition to a green global financial system
The impact of institutional distance on the sustainable development of countries involved in China's Belt and Road initiative	The challenges and implications of ESG evaluations and sustainable investment behavior	
Evaluation of the resilience effectiveness of green bonds and ESG investments against economic and climate shocks	Determinants of willingness to pay for sustainable investments, including investor preferences and product certification	

4 Discussion

The discussion surrounding sustainable investment in the contexts of the China, Germany and United States provides valuable insights into the global strategies and local challenges that shape the dynamics of global financial markets. Each country, with its own policies, economic priorities, and environmental commitments, offers a unique perspective on the evolution of sustainable finance. This comparative analysis highlights not only the differences and similarities between these regions but also the distinctive aspects that could influence future trends in sustainable investment on a global scale.

4.1 Similarities in Terms of Global Approach and Transnational Synergies

Research shows a convergence towards the adoption of ESG criteria as the foundation of sustainable investment policies. This phenomenon is particularly evident in recent studies from the United States, where researchers such as [17] highlight how global initiatives like the SDGs interact with national policies to bolster sustainable investments. Similarly, in Germany and China, the literature reveals a growing interest in integrating ESG criteria into financial decision-making, indicating a global perspective on responsible investment [18,19]. In all three regions, the implementation of strict regulations concerning ESG investments is seen as crucial for ensuring transparency and reliability in investment practices. In the United States, the implications of major initiatives such as the SDGs are examined for their ability to shape sustainable investment practices, while Germany and China focus more on harmonizing ESG standards to promote a fair transition towards a green economy [20].

4.2 Divergences Based on Local Contexts

The United States shows a trend of integrating sustainable development initiatives within a broader economic policy framework, highlighting challenges such as barriers to adopting green technologies [21]. However, [22] contributed to the literature on the response of financial markets and investment decisions in the context of the COVID-19 pandemic. In Germany, the resilience of ESG indices during the COVID-19 crisis underscored the challenges of financial performance in sustainable investments, raising questions about the stability of these instruments during times of economic turbulence [23]. In addition to economic factors, psychological factors also played a decisive role in investors' decisions, during the COVID-19 pandemic, behavioral responses to fear, anxiety regarding market volatility, risk propensity, and herd behavior influenced the decision-making of socially responsible investors [24]. Meanwhile, China focuses on divergences in ESG ratings and their impact on financial stability, reflecting concerns about the consistency and effectiveness of ESG assessments in investment decisions [25]. Cultural and economic specificities also significantly influence each country's approach to sustainable investment. For example, German studies highlight the role of financial education and cultural preferences in shaping attitudes towards sustainable investments [26]. In China, the emphasis is on aligning investment strategies with the country's sustainable development policies, particularly with respect to green infrastructure investments under the Belt and Road Initiative [27].

5 Conclusion

The comparative analysis of sustainable investment practices in the United States, Germany, and China underscores a global shift toward integrating ESG criteria into financial decision-making. This convergence aligns with the SDGs, reflecting the growing importance of sustainable finance in addressing global challenges. However, regional disparities persist: the United States faces barriers in green technology adoption, Germany encounters ESG index vulnerabilities during economic shocks, and China struggles with inconsistencies in ESG ratings, threatening financial stability. To address these challenges, advanced machine learning methodologies, such as those proposed by [28–30], offer a promising approach to enhance the precision and scalability of ESG analysis. Additionally, incorporating contextual factors, as explored by [31], can improve the understanding and implementation of sustainable investment strategies. By integrating these methodological and contextual insights, future research can support the development of harmonized ESG standards and localized strategies, advancing global sustainability objectives.

References

1. Ahmad, H., Yaqub, M., Lee, S.H.: Environmental-, social-, and governance-related factors for business investment and sustainability: a scientometric review of global trends. Environ. Dev. Sustain. **26**(2), 2965–2987 (2024)
2. Goutte, S., Sanin, M.-E.: Main challenges regarding development and sustainability in economics and finance. Dev. Sustain. Econ. Fin., 100002 (2024)
3. Dimmelmeier, A.: Expanding the politics of measurement in sustainable finance: reconceptualizing environmental, social and governance information as infrastructure. Environ. Plann. C Polit. Space **42**(5), 761–781 (2024)
4. Lehner, O.M.: Routledge Handbook of Social and Sustainable Finance. Routledge (2016)
5. Sampieri, S., Saoualih, A., Safaa, L., Carnero Calzada, F.M., Ramazzotti, M., Martínez-Peláez, A.: Tourism development through the sense of UNESCO world heritage: the case of Hegra, Saudi Arabia. Heritage **7**(4), 2195–2216 (2024)
6. Saoualih, A., et al.: Exploring the tourist experience of the Majorelle garden using VADER-based sentiment analysis and the latent Dirichlet allocation algorithm: the case of TripAdvisor reviews. Sustainability **16**(15), 6378 (2024)
7. Beriozovas, O., Perkumienė, D., Škėma, M., Saoualih, A., Safaa, L., Aleinikovas, M.: Research advancement in forest property rights: a thematic review over half a decade using natural language processing. Sustainability **16**(19), 8280 (2024)
8. Traag, V.A., Waltman, L., Van Eck, N.J.: From Louvain to Leiden: guaranteeing well-connected communities. Sci. Rep. **9**(1), 1–12 (2019)
9. Blondel, V.D., Guillaume, J.-L., Lambiotte, R., Lefebvre, E.: Fast unfolding of communities in large networks. J. Stat. Mech. Theory Exp. **2008**(10), 10008 (2008)
10. Girvan, N., Girvan, M., Newman, M.E.J.: Community structure in social and biological networks. Proc. Nat. Acad. Sci. **99**(12), 7821–7826 (2002)
11. Newman, M.E.: Finding community structure in networks using the eigenvectors of matrices. Phys. Rev. E-Stat. Nonlinear Soft Matter Phys. **74**(3), 036104 (2006)
12. Clauset, A., Newman, M.E., Moore, C.: Finding community structure in very large networks. Phys. Rev. E-Stat. Nonlinear Soft Matter Phys. **70**(6), 066111 (2004)
13. Raghavan, U.N., Albert, R., Kumara, S.: Near linear time algorithm to detect community structures in large-scale networks. Phys. Rev. E-Stat. Nonlinear. Soft Matter Phys. **76**(3), 036106 (2007)

14. Pons, P., Latapy, M.: Computing communities in large networks using random walks. In: Yolum, I., Güngör, T., Gürgen, F., Özturan, C. (eds.) ISCIS 2005. LNCS, vol. 3733, pp. 284–293. Springer, Heidelberg (2005). https://doi.org/10.1007/11569596_31
15. Rosvall, M., Bergstrom, C.T.: Maps of random walks on complex networks reveal community structure. Proc. Natl. Acad. Sci. **105**(4), 1118–1123 (2008)
16. Reichardt, J., Bornholdt, S.: Statistical mechanics of community detection. Phys. Rev. E-Stat. Nonlinear Soft Matter Phys. **74**(1), 016110 (2006)
17. Valentini, N., Jarrah, R., Chen, C.-R.: Public perception of residential solar energy in Minnesota's urban areas. In: Economics and Policy of Energy and the Environment (2023/1) (2023). FrancoAngeli Editore
18. Gutsche, G., Nakai, M., Arimura, T.H.: Revisiting the determinants of individual sustainable investment-the case of Japan. J. Behav. Exp. Fin. **30**, 100497 (2021)
19. Luo, D., Shan, X., Yan, J., Yan, Q.: Sustainable investment under ESG volatility and ambiguity. Econ. Model. **128**, 106471 (2023)
20. Brühl, V.: Green financial products in the EU-a critical review of the status quo. Intereconomics **57**(4), 252–259 (2022). https://doi.org/10.1007/s10272-022-1057-2
21. Kölbel, J.F., Heeb, F., Paetzold, F., Busch, T.: Can sustainable investing save the world? Reviewing the mechanisms of investor impact. Organ. Environ. **33**(4), 554–574 (2020). https://doi.org/10.1177/1086026620919202
22. Kartobi, S.E., Oubida, A.A.: The impact of the COVID-19 pandemic on stock exchanges. In: Public Health and Economic Resiliency in the Post-COVID-19 Era, pp. 156–168. IGI Global (2022). https://www.igi-global.com/chapter/the-impact-of-the-covid-19-pandemic-on-stock-exchanges/287034
23. Schumacher, K., Chenet, H., Volz, U.: Sustainable finance in Japan. J. Sustain. Fin. Investment **10**(2), 213–246 (2020). https://doi.org/10.1080/20430795.2020.1735219
24. Amine, O.M., Zineb, E., Eddine, K.S.: Study of the behavioural determinants of investment in the era of the COVID-19 pandemic among socially responsible investors in Morocco. Economic Archive/Narodnostopanski Arhiv, vol. 2 (2023)
25. Liu, Y., Wang, J., Wen, F., Wu, C.: Climate policy uncertainty and bank systemic risk: a creative destruction perspective. J. Fin. Stab., 101289 (2024)
26. Zwergel, B., Wins, A., Klein, C.: On the heterogeneity of sustainable and responsible investors. J. Sustain. Fin. Investment **9**(4), 282–294 (2019). https://doi.org/10.1080/20430795.2019.1613820
27. Pan, C., Sun, T., Mirza, N., Huang, Y.: The pricing of low emission transitions: evidence from stock returns of natural resource firms in the GCC. Resour. Pol. **79**, 102986 (2022)
28. Lyaqini, S., Hadri, A., Ellahyani, A., Nachaoui, M.: Primal dual algorithm for solving the Nonsmooth twin SVM. Eng. Appl. Artif. Intell. **128**, 107567 (2024)
29. Nachaoui, M., Lyaqini, S., Chaouch, M.: Indicating if water is safe for human consumption using an enhanced machine learning approach. Stat. Optim. Inf. Comput. **11**(1), 70–81 (2023)
30. Lyaqini, S., Nachaoui, M.: Diabetes prediction using an improved machine learning approach. Math. Model. Comput. **8**(1), 726–735 (2021)
31. Khoudi, Z., Nachaoui, M., Lyaqini, S.: Identifying the contextual factors related to the reading performance of Moroccan fourth-grade students from a machine learning-based approach. Educ. Inf. Technol. **29**(3), 3047–3073 (2024)

Masked Face Recognition: A Comprehensive Review of Techniques, and Datasets

Siham Ahmam(✉), Nidal Lamghari, and Abdelghani Ghazdali

LIPIM, ENSA Khouribga, Khouribga, Morocco
siham.ahmam@usms.ac.ma, {n.lamghari,a.ghazdali}@usms.ma

Abstract. Facial recognition is essential for identification and identity verification, especially used in security and authentication. The COVID-19 pandemic posed a major challenge due to the widespread use of masks, compromising the effectiveness of facial recognition systems. This survey presents a thorough examination of the advancements in masked face recognition. We explore various techniques, including feature extraction from visible facial regions, and the reconstruction of occluded areas using methods such as generative models and transfer learning. This survey aims to provide valuable insights for overcoming occlusion challenges and advancing the development of effective biometric systems for recognizing masked faces.

Keywords: Masked face recognition · Face detection · Convolutional Neural network · Generative Adversarial Network · Transfer learning

1 Introduction

Facial recognition technology analyses unique facial characteristics to authenticate or establish a person's identity. It is frequently used in security systems, cell phones, and other authentication and access control applications. In an era of technical advancements and broad integration of artificial intelligence, face recognition emerges as a critical frontier. However, the advent of unanticipated conditions, such as the global acceptance of face masks, has posed fundamental and significant problems to this profession. As the globe grapples with the COVID-19 epidemic and the need for public health measures, the prevalence of masks has radically impacted human relationships and cultural norms. Despite this paradigm change, reliance on face recognition technology remains, necessitating inventive ways to adapt to the new reality. Occlusion, or covering a portion of a person's face, has long been a worry for facial recognition. It makes identification harder. Despite advancements in occlusion reduction, facial recognition systems still have difficulties. The COVID-19 outbreak has sparked curiosity in

N. Lamghari, A. Ghazdali—Contributed equally to this work.

A. Laghrib and A. Ghazdali (Eds.): ICMDS 2024, LNNS 1466, pp. 203–212, 2025.
https://doi.org/10.1007/978-3-031-94802-2_20

developing face recognition software that can recognize mask-wearing individuals. Among the challenges include the lack of a masked face dataset, different mask designs, privacy issues, and the challenge of facial feature retrieval due to mask coverage. Because masks obscure important face traits, they have made facial recognition systems' work much more difficult. The amount and quality of information available for identification are significantly reduced when facial features including the lips, nose, and portion of the cheeks are hidden. As a result, identifying and validating identities is extremely challenging for traditional facial recognition algorithms, which mostly depend on these traits for face identification.

2 Face Recognition

2.1 Face Recognition Datasets

Several publicly available datasets have been instrumental in advancing research on face recognition, particularly under diverse and unconstrained conditions. The IMDb-Face dataset [15] is a comprehensive collection built for face recognition research, comprising roughly 1.7 million faces and 59,000 identities. Labeled Faces in the Wild (LFW) [14] is a series of face images developed primarily to investigate the difficulty of unrestricted face recognition. It has approximately 13,000 face photos collected from the internet, each with the name of the person depicted. 1,680 individuals in the sample have at least two different images. The CelebFacesAttributes-Dataset (CelebA) [16] contains more than 200,000 celebrity pictures, each with 40 face-related attribute annotations. The dataset includes images with diverse poses and background settings. CelebA is characterized by its wide range of attributes, a large volume of data, and comprehensive annotations. The MegaFace dataset [17] contains 4,753,320 faces of 672,057 identities from 3,311,471 photos sourced from 48,383 Flickr users' albums. The UMDFaces dataset [18] is split into two sections: Still Images and Video Frames. The Still Images part includes 367,888 face annotations for 8,277 subjects categorized into 3 batches. These annotations encompass human-curated bounding boxes for faces, estimated pose (yaw, pitch, and roll), and keypoint locations. VGGFace2 [19] contains around 3.31 million pictures divided into 9,131 classes. The dataset is divided into training and testing parts, with 170,000 photographs in the test set and the remainder in training. CASIA-WebFace [21] is employed for face verification and face identification tasks. The dataset comprises 494,414 face images of 10,575 real identities, created by the Institute of Automation, Chinese Academy of Sciences (CASIA). Lastly, the Extended Yale B database [23] comprises 2,414 frontal-face images measuring 192×168 pixels, featuring 38 subjects with approximately 64 images per subject. These images were taken under diverse lighting conditions and with different facial expressions.

A summary of the face datasets used in face recognition research, as discussed above, is presented in Table 1.

Table 1. Comparative Evaluation of Face Recognition Datasets: Type, Image Count, and Subjects

Dataset	Type	Images	Subjects
IMDBB-Face	Unmasked	1.7 million	59,000
LFW	Unmasked	13,000	1,680
CelebFaces	Unmasked	202,599	10,177
MegaFace	Unmasked	4,753,320	672,057
UMDFaces	Unmasked	367,888 images	8,277
		3,735,476 videos	3,107
VGGFace2	Unmasked	3.31 million	9,131
CASIA-WebFace	Unmasked	494,414	10,575
Extend Yale B	Unmasked	2,414	38

2.2 Face Recognition Techniques

Facial recognition has become critical in a variety of industries, including security and surveillance, as well as creating individualized user experiences. Facial recognition algorithms have improved in accuracy and resilience as a result of advances in machine learning and computer vision, especially under difficult settings such as occlusion, shifting illumination, and varied positions. This progress has been made possible by innovative techniques and models that constantly push the boundaries of what is achievable in this discipline. Boutros et al. [27] proposed the use of elastic penalty margin loss (ElasticFace) to relax the fixed penalty margin constraint, allowing for more flexibility in driving class separability. Wang et al. [28] introduced the large margin Cosine loss (LMCL) as a novel loss function for leveraging deep CNNs to acquire more distinctive features for face recognition. They transformed the Softmax loss into a cosine loss by utilizing L2-normalized weights and feature vectors to mitigate radial fluctuations. Hayat et al. [29] presented a novel hybrid loss function based on an affinity measure in Euclidean space to address class unbalanced learning using deep CNN. The goal is to reduce intra-class variability and maximize inter-class distances. Experimental results showed the efficiency of this loss function for face verification across multiple datasets. Sun et al. [30] introduced inter-class angular margin (IAM) loss to dynamically expand inter-class diversity by imposing greater penalties on smaller inter-class angles. This increases the angular margin between classes and enhances facial feature discrimination. In 2019, Song et al. [31] developed a PDSN architecture to detect the relationship between corrupted feature components and occluded facial blocks in deep CNN models. Prakash et al. [32] proposed a system combining CNN and transfer learning for automated face recognition. To train the face database images, a CNN was utilized with weights obtained from the pre-trained VGG-16 model on the extensive ImageNet database. The extracted features were input into the fully connected layer with softmax activation for classification. Khan et al. [33] described a system for facial recognition using transfer learning and pre-trained neural networks. AlexNet, a pre-trained convolutional neural network, was utilized and achieved an accuracy of 97.95%. Benradi et al. [34] suggested a novel method to enhance face recognition accuracy under variation or occlusion by integrating feature extraction techniques such as histograms of oriented gradient (HOG), SIFT, Gabor, and Canny contour detectors, along with a CNN architecture. Finally, Hangaragi et al. [35] proposed a model utilizing Face Mesh for face identification and recognition to prevent unauthorized access to restricted areas. This model demonstrated robust performance in diverse environments, achieving a face recognition accuracy of 94.23% when trained on LFW and real-time images.

3 Masked Face Recognition

3.1 Masked Face Recognition Datasets

Various specialized datasets have been introduced to facilitate research and development in the field of masked face recognition. The following provides a summary of the most prominent datasets in this area:

The MaskedFace-Net dataset [13] was created primarily for training and testing face recognition algorithms under masked situations. It was intended to solve the issues raised by face masks, which conceal crucial facial characteristics and make it difficult for face recognition systems to function successfully. The dataset is a synthetic expansion of the CelebA dataset. The original images were digitally enhanced using masks created using generative methods. There are around 137,000 photos, with each variant (Correctly Masked Faces and Incorrectly Masked Faces) including approximately 68,000 images. The Artificially-Simulated-Masked Face Dataset (ASMFD) [12] includes both faces and masked faces, detected and cropped by YOLOv3 from publicly available datasets and videos. The cropped face images come in various sizes and resolutions, labelled as either masked or no-mask faces. The RMFD (Real-World Masked Face Dataset) [20] contains 5,000 anonymized images of 525 individuals and 90,000 regular images. The CASIA-WebFace-Mask dataset [22] incorporates synthetic face masks into the images to support tasks related to recognizing masked faces, which has become increasingly important due to the widespread use of face masks. To simulate real-world conditions, various studies have created masked versions of CASIA-WebFace by digitally adding masks. The LFW-SMFD dataset [24] is derived from the Labeled Faces in the Wild (LFW) dataset, which contains images of well-known individuals gathered from a website. Within the LFW dataset, the SMFRD also contains the same images, but the individuals' faces are masked. There are 13,117 images of 5,713 individuals in the dataset. The training set includes 13,027 masked images of 5,713 individuals, while the test set includes 70 masked images of 48 individuals. The COMASK20 dataset [25] contains 2,754 pictures of individuals wearing masks, featuring 300 different people. These pictures are taken from video footage, providing a range of angles and situations to assist with activities like identifying and detecting masked faces. Lastly, the MFRD-80K dataset [26] contains 80,500 images of masked faces from 161 individuals in real-life scenarios.

The comparison between these datasets is summarized in Table 2.

Table 2. Comparative Evaluation of Masked Face Recognition Datasets: Type, Image Count, and Subjects

Dataset	Type	Images	Subjects
ASMFD	Masked and No Masked	835 masked, 1,227 no-mask	1,227
RMFD	Masked	5,000	525
CASIA-WebFace-Mask	masked	445,446	10,575
LFW-SMFD	Masked	13,117	5,713
COMASK20	Masked and No Masked	2,754	300
MFRD-80K	Masked	80,500	161

3.2 Masked Face Recognition Techniques

Recent progress in masked facial recognition can be attributed to notable developments in computer vision, deep learning, and image processing technologies. Researchers and practitioners have worked to tackle the specific challenges posed by masked faces, especially the diminished visibility of facial features. Wang et al. [1] suggested enhancing Facenet by integrating the ECA (Efficient Channel Attention) mechanism with ConvNeXt-T to extract more valuable information from the unobscured area of the face. This approach increases attention to key areas in the feature map without raising the model's overall complexity. Eman et al. [2] proposed integrating resilient principal component analysis (RPCA), oval and landmark face detection, and deep learning-based mask detection to present a novel approach to masked face recognition. They employed a pre-trained SSD-MobileNetV2 model to locate and identify masks, using landmark and oval face detection to identify key facial characteristics. Malakar et al. [4] introduced a technique that augments only the lower portion of masked faces rather than reconstructing the entire face. This approach enhances SSIM (Structural Similarity Index Measure) and improves recognition accuracy. Transfer learning, a machine learning technique, enables models trained for one task to be adapted for another. Shukla et al. [3] fine-tuned a pre-trained MobileNet V2 model for masked face detection, using transfer learning to minimize the need for large amounts of training data. Mosleh et al. [7] tackled the issue of occlusion caused by masks by employing transfer learning models to enhance facial recognition accuracy in masked environments. Liu [8] presented a lightweight mask-wearing recognition method built on MobileNetV2, achieving 98.30% accuracy and making it suitable for mobile devices. Ramgopal et al. [9] fine-tuned ResNet-50 for masked datasets, improving recognition performance and processing efficiency, especially in surveillance systems. To address the challenges of removing unwanted objects from images, two major issues must be tackled: detecting and removing the object, and seamlessly completing the image. Maimuna et al. [11] utilized GANs to reconstruct missing facial features masked by masks, enhancing perceptual quality for improved recognition outcomes. Farahanipad et al. [5] proposed a GAN-based unpaired domain translation method to transform masked face images into unmasked ones. Boutros et al. [6] suggested an embedding unmasking model that uses masked face embeddings to generate embeddings resembling unmasked faces. Agarwal et al. [10] introduced a method combining GANs and residual learning to restore masked facial features, achieving superior recognition results.

The comparison between these datasets is summarized in Table 2.

4 A Comparative Study

In this section, we present a comparative study of various approaches proposed for both masked and unmasked facial recognition. The comparison focuses on

Table 3. Comparative Study of Face Recognition Approaches: Methods, Datasets, and Accuracy Performance.

Approch	Method	Dataset	Accuracy
Boutros et al. [27]	Elastic Loss	LFW	99.82%
		AgeDB-30	98.35%
		CPLFW	93.28%
		CFP-FP	95.43%
		IJB-C	98.92%
Wang et al. [28]	large margin cosine loss (LMCL)	LFW	99.73%
	loss (LMCL)	YTF	97.6%
Hayat et al. [29]	Affinity Loss	LFW	99.65%
		YTF	97.3%
		AgeDB30	95.9%
		CFP-FP	96.0%
Sun et al. [30]	Inter-class angular	LFW	99.2%
	margin (IAM) loss	YTF	95.52%
		CFP	93.34%
Song et al. [31]	Pairwise differential	LFW	99.20%
	siamese network (PDSN)	MegaFace	74.40%
		AR	99.72%
Prakash et al. [32]	CNN with Transfer learning	Yale	98.7%
		AT&T	98.7%
Khan et al. [33]	AlexNet with Transfer learning	propre dataset	97.95%
Benradi et al. [34]	SIFT+ CNN	ORL and Sheffield	100%
Benradi et al. [35]	Face Mesh+ Deep NN	LFW	94.23%

the methods employed, the datasets utilized, and the accuracy achieved by each approach.

Table 3 highlights the comparison of methods specifically designed for unmasked facial recognition, detailing the techniques used, the databases tested, and their corresponding performance metrics.

Table 3 highlights the diverse methods used in face recognition, showcasing varying levels of performance across different datasets. Approaches like Boutros et al.'s Elastic Loss [27] and Wang et al.'s LMCL [28] achieve high accuracy, particularly on popular datasets such as LFW, demonstrating the effectiveness of advanced loss functions in improving recognition accuracy. However, some methods, like Song et al.'s PDSN [31], show a performance drop on larger datasets like MegaFace, indicating the challenges of scaling certain techniques. The table also emphasizes the potential of transfer learning, as seen by Prakash et al. [32] and Khan et al. [33], which delivers strong results with limited data.

Table 4 summarizes the different methods cited in the state of the art on masked facial recognition. It shows a comparison between these methods according to the method, the dataset and the accuracy obtained by this method.

Table 4. Comparative Study of Masked Face Recognition Approaches: Methods, Datasets, and Accuracy Performance.

Approch	Method	Dataset	Accuracy
Wang et al. [1]	ConvNeXt-T+ ECA	CASIA-WebFace and LFW	99.76%
Eman et al. [2]	RPCA +PSO+ KNN	LFW-SMFD	97%
Shukla et al. [3]	Transfer learning using MobileNet V2	CIFAR10	92.4%
		MNIST	93.9%
		RMFRD	70%
		SMFRD	69%
		Thier proposed dataset	99.82%
Malakar et al. [4]	SURF + CNN	LFW	98%
		CASIA-WebFace	91.3%
		AR	96%
		FACES datasets	93.3%
Farahanipad et al. [5]	GAN	-	-
Boutros et al. [6]	EUM + Self-restrained Triplet (SRT)	-	-
Mosleh et al. [7]	pre-trained DCNN	CASIA-WebFace mask	93%
Liu et al. [8]	MobileNetV2	Masked face-net dataset	98.30%
Ramgopal et al. [9]	YOLOv3	RWMFD	-
Agarwal et al. [10]	GAN	-	-
Maimuna et al. [11]	GAN	-	-

The comparison of different approaches for masked face recognition, as shown in Table 4, highlights the wide range of methods used to tackle the challenges posed by occlusion in face recognition tasks. The accuracy performance varies significantly depending on the method, dataset, and specific challenges addressed by each approach. While some methods achieve high accuracy on standard datasets like CASIA-WebFace and LFW, others, especially those leveraging advanced techniques such as GANs or transfer learning, demonstrate potential but show varied success depending on the specific conditions and datasets. This table emphasizes the diverse range of strategies in the field and indicates the ongoing need for further advancements to enhance the robustness and applicability of masked face recognition systems in real-world settings.

5 Conclusion

This research covers a comprehensive evaluation of the latest advanced methods for recognizing both masked and unmasked faces, as well as the various datasets used for training and assessment. As the use of face masks continues to increase, accurately identifying masked faces has become more challenging. While unmasked face recognition relies on the complete display of facial features, recognizing masked faces presents unique challenges due to the covering of crucial areas. Therefore, techniques and adaptations are necessary to effectively address partial face visibility and ensure accurate identification despite these limitations.

References

1. Wang, Y., Li, Y., Zou, H.: Masked face recognition system based on attention mechanism. Information **14**, 87 (2023)
2. Eman, M., Mahmoud, T.M., Ibrahim, M.M., Abd El-Hafeez, T.: Innovative hybrid approach for masked face recognition using pretrained mask detection and segmentation, robust PCA, and KNN classifier. Sensors **23**, 6727 (2023)
3. Shukla, R.K., Tiwari, A.K.: Masked face recognition using MobileNet v2 with transfer learning. Comput. Syst. Sci. Eng. **45**(1), 293–309 (2023)
4. Malakar, S., Chiracharit, W., Chamnongthai, K.: Masked face recognition with generated occluded part using image augmentation and CNN maintaining face identity. IEEE Access **12**, 126356–126375 (2024)
5. Farahanipad, F., Rezaei, M., Nasr, M., Kamangar, F., Athitsos, V.: GAN-based face reconstruction for masked-face (2022)
6. Boutros, F., Damer, N., Kirchbuchner, F., Kuijper, A.: Unmasking face embeddings by self-restrained triplet loss for accurate masked face recognition. arXiv preprint arXiv:2103.01716 (2021)
7. Mosleh, M.A.A., et al.: Masked face recognition using transfer learning approaches. In: ICESP Signals and Communication Technology. Springer, Cham (2023)
8. Liu, M.: Masked face recognition based on transfer learning, Doctoral dissertation, Auckland University of Technology (2022)
9. Ramgopal, M., Roopesh, M.S., Chowdary, M.V., et al.: Masked facial recognition in security systems using transfer learning. SN Comput. Sci. **4**, 27 (2023)

10. Agarwal, C., Bhatnagar, C., Mishra, A.: Reconstructing masked face using GAN technique. In: ICAISA. LNCS, vol. 756. Springer, Singapore (2023)
11. Maimuna, M., Hossain, S., Deb, K.: Masked face inpainting using generative adversarial network based architecture. In: ICO. LNNS, vol. 569. Springer, Cham (2022)
12. Boutros, F., Naser, D., Florian, K.: Artificially simulated masked face dataset. Mendeley Data, V1 (2021)
13. Cabani, A., Hammoudi, K., Benhabiles, H., Melkemi, M.: MaskedFace-Net–a dataset of correctly/incorrectly masked face images in the context of COVID-19. Smart Health **19**(100144), 2352–6483 (2021)
14. Huang, G.B., Mattar, M., Berg, T., Learned-Miller, E.: Labeled faces in the wild: a database for studying face recognition in unconstrained environments. In: Workshop on Faces in Real-Life Images: Detection, Alignment, and Recognition (2008)
15. Wang, J., Zhu, X., Gong, S., Li, W.: IMDb-Face: a large-scale dataset for face recognition in the wild. arXiv preprint arXiv:1808.05008 (2018)
16. Liu, Z., Luo, P., Wang, X., Tang, X. Deep learning face attributes in the wild. In: IEEE ICCV, pp. 3730–3738 (2015)
17. Kemelmacher-Shlizerman, I., Matusik, W., Shih, M.: The MegaFace benchmark: 1 million faces for recognition at scale. IEEE TPAMI **38**(9), 1798–1808 (2016)
18. Bansal, A., Nanduri, A., Castillo, C.D., Ranjan, R., Chellappa, R.: UMDFaces: an annotated face dataset for training deep networks. In: IJCB, pp. 464–473 (2017)
19. Cao, Q., Shen, L., Xie, W., Parkhi, O.M., Zisserman, A. VGGFace2: a dataset for recognising faces across pose and age. In: IEEE FG, pp. 67–74 (2018)
20. Wang, Z., Huang, B., Wang, G., Yi, P., Jiang, K.: Masked face recognition dataset and application. IEEE T-BIOM **5**(2), 298–304 (2023)
21. Yi, D., Lei, Z., Liao, S., Li, S.Z.: Learning face representation from scratch. arXiv preprint arXiv:1411.7923 (2014)
22. Mare, T., et al.: A realistic approach to generate masked faces applied on two novel masked face recognition data sets (2021)
23. Georghiades, A.S., Belhumeur, P.N., Kriegman, D.J.: From few to many: illumination cone models for face recognition under variable lighting and pose. IEEE TPAMI **23**(6), 643–660 (2001)
24. Anwar, A., Raychowdhury, A.: Masked face recognition for secure authentication. arXiv preprint arXiv:2008.11104 (2020)
25. Vu, H.N., Nguyen, M.H., Pham, C.: Masked face recognition with convolutional neural networks and local binary patterns. Appl. Intell. **52**, 5497–5512 (2022)
26. Lee, C.P., Lim, K.M.: MFRD-80K: a dataset and benchmark for masked face recognition. Eng. Lett. **29**(4), 1595–1600 (2021)
27. Boutros, F., Damer, N., Kirchbuchner, F., Kuijper, A.: ElasticFace: elastic margin loss for deep face recognition. In: CVPR, pp. 1578–1587 (2022)
28. Wang, H., Wang, Y., Zhou, Z., Ji, X., Gong, D., Zhou, J., Liu, W.: CosFace: large margin cosine loss for deep face recognition. In: CVPR, pp. 5265–5274 (2018)
29. Hayat, M., Khan, S., Zamir, W., Shen, J., Shao, L.: Max-margin class imbalanced learning with gaussian affinity. arXiv preprint (2019)
30. Sun, J., Yang, W., Gao, R., Xue, J.-H., Liao, Q.: Inter-class angular margin loss for face recognition. Sig. Process. Image Commun. **80**, 115636 (2020)
31. Song, L., Gong, D., Li, Z., Liu, C., Liu, W.: Occlusion-robust face recognition based on mask learning with pairwise differential Siamese network. In: Proceedings of the IEEE/CVF International Conference on Computer Vision, pp. 773–782 (2019)
32. Prakash, R.M., Thenmoezhi, N., Gayathri, M.: Face recognition with convolutional neural network and transfer learning. In: ICSSIT, pp. 861–864. IEEE (2019)

33. Khan, S., Ahmed, E., Javed, M. H., Shah, S. A., Ali, S. U.: Transfer learning of a neural network using deep learning to perform face recognition. In: ICECCE, pp. 1–5. IEEE (2019)
34. Benradi, H., Chater, A., Lasfar, A.: A hybrid approach for face recognition using a convolutional neural network combined with feature extraction techniques. IAES Int. J. Artif. Intell. **12**, 627–640 (2023)
35. Hangaragi, S., Singh, T., Neelima, N.: Face detection and recognition using face mesh and deep neural network. Procedia Comput. Sci. **218**, 741–749 (2023)

Lightweight Real-Time IDS for WSNs Through Dimensionality Reduction and Ensemble Learning

Hiba Tabbaa(✉), Mourad Jabrane, and Imad Hafidi

Laboratory of Process Engineering, Computer Science and Mathematics (LIPIM), National School of Applied Sciences Khouribga, University Sultan Moulay Slimane, Khouribga, Morocco

{hiba.tabbaa,mourad.jabrane}@usms.ac.ma, i.hafidi@usms.ma

Abstract. Wireless Sensor Networks (WSNs) are widely deployed across many critical infrastructures because of their low cost, flexibility, and efficiency, besides their broad applicability across various domains. A normal WSN is basically composed of small, battery-powered nodes that monitor environmental conditions and endorse real-time applications; thus, WSNs are very prone to security threats, especially Denial of Service (DoS) attacks that disrupt network functionality. The traditional security mechanisms tend to fall short, as WSNs lack the computational, memory, and energy resources for complex classification algorithms. In this regard, we aim to develop a resilient Intrusion Detection System (IDS) optimized for WSNs, capable of real-time detection and mitigation against security breaches. Our IDS leverages an online ensemble learning approach based on the Adaptive Random Forest with Hoeffding Adaptive Tree (ARF+HAT) algorithm, alongside three dimensionality reduction (DR) techniques: Gaussian Random Projection (GRP), Random Bernoulli Projection (RBP), and Hashing Trick (HT). The IDS operates in three online phases: pre-processing, dimensionality reduction, and attack detection, where DR selectively retains the most relevant features to reduce computational demands. Extensive simulations validate the IDS's effectiveness, with HT combined with the ARF+HAT achieving high accuracy, precision, recall, and F1-score, while offering superior memory and runtime efficiency. HT's low false alarm rate and enhanced energy efficiency underscore its suitability for real-time intrusion detection in resource-constrained WSN environments, positioning it as a practical and robust solution for improving WSN security.

Keywords: Wireless Sensor Network (WSN) · Intrusion Detection System (IDS) · DoS Attack · Security · Machine Learning (ML) · Online Learning (OL) · Dimensionality Reduction · Ensemble Learning (EL) · Data Streams

A. Laghrib and A. Ghazdali (Eds.): ICMDS 2024, LNNS 1466, pp. 213–225, 2025.
https://doi.org/10.1007/978-3-031-94802-2_21

1 Introduction

Over recent years, with the constant growth of the Internet and rapid technological advances in microelectronics and wireless communication technologies, a new variant of ad hoc networks has emerged: Wireless Sensor Networks (WSNs). WSNs have brought forth a diverse array of applications and have gained prominence in current research. As information technology progressed, WSNs have also become a fundamental part of the Internet of Things (IoT) [1].

The WSNs consist of a collection of independent sensor nodes spatially scattered across an area of interest. Each node is equipped with a sensing unit that monitors various physical phenomena, such as heat, humidity, pressure, temperature, vibrations, noise levels, and converts them into digital data. Nodes also contain a processing unit, data storage, a wireless transmission module, and a battery, enabling them to transmit the collected sensory data to a base station [2]. WSNs utilize various types of sensors, such as thermal, visual, radar, infrared, and magnetic, enhancing their versatility across different environments, including extreme conditions where traditional wired and wireless networks (e.g. WLAN) may be impractical [3,4].

Security is essential for WSN applications, particularly when sensor nodes are deployed in insecure locations like battlefields or critical infrastructure. These unprotected, hard-to-reach nodes, often without battery recharging options, are vulnerable to malicious actions that threaten the core functionality of WSN. Since the primary purpose of WSN nodes is to collect monitoring data and transmit it to an end-user for analysis, interpretation, and decision-making, this process must be done without malicious interference and with a strong level of security [5–7].

Traditional security mechanisms often fail short, as WSNs lack the computational, memory, and energy resources needed for complex classification algorithms. The foremost challenge in developing a robust Intrusion Detection System (IDS) for WSNs lies in achieving high accuracy in attack detection while addressing constraints essential to extending system longevity and reliability. Given the impracticality of recharging or replacing nodes in many deployments, particularly in remote monitoring, lightweight, energy-efficient detection methods are crucial for enhancing network lifespan.

Evidently, this objective can be approached through various ways, particularly by refining detection strategies and Dimensionality Reduction (DR) techniques employed. Doing so reduces the overhead introduced by online IDS mechanisms and optimizes resource consumption without compromising performance [8]. Data DR arises as a key factor in transforming and selecting only the most relevant features from WSN data streams in order to reduce algorithm space and time demands. By removing redundancies and isolating critical features, DR optimizes resource usage, facilitates classification, and improves data visualization, making complex data patterns easier to interpret. Beyond any doubt, this approach reduces the overhead introduced by real-time IDS mechanisms in securing WSN environments, enhancing resource efficiency without compromising performance.

The structure of this paper is as follows: Sect. 2 provides an overview of the key related works. Section 3 outlines the methodology of our proposed approach. Section 4 presents the experimental results. Finally, conclusions are drawn in Sect. 5.

2 Related Works

To protect WSNs from malicious attacks, several IDS strategies have been introduced in the literature, primarily categorized into anomaly detection (AD), misuse detection, specification-based detection, and hybrid detection systems [9]. Given its ability to identify both known and novel attacks, anomaly-based IDS have received considerable attention, with approaches ranging from supervised to unsupervised and hybrid models [10]. Our work builds on this foundation, focusing specifically on the application of anomaly-based IDS to streaming data in WSNs, a relatively under-explored area in current literature.

A large body of theoretical and empirical research has shown that combining the detecting results of several individual detectors can drastically enhance efficiency [11,12], but the original Ensemble Learning (EL) method typically necessitates the construction and storage of several individual detectors, which necessitates a significant amount of computing and storage resources and may not be suitable for WSNs. One approach to address this limitation is the application of DR techniques to reduce the computational and storage burden while preserving detection performance.

Few studies have proposed optimized ensemble tailored to the needs of IoT settings. Martindale et al. [13] highlighted the trade-offs between accuracy and resource demands in online ensemble learning, finding that a heterogeneous ensemble of Adaptive Random Forest (ARF) and Hoeffding Adaptive Tree (HAT) outperformed individual models at a higher computational cost. Similarly, Ding et al. [14] proposed an optimized online ensemble approach using biogeographical-based optimization (BBO) for pruning, achieving effective detection on a WSN dataset with reduced resource usage.

To the best of our knowledge, this is the first work to apply DR to an online ensemble learning model composed of ARF and HAT specifically for AD in WSNs, addressing both performance and resource constraints in high-dimensional streaming data.

3 Methodology

The infinite and evolving nature of data streams in WSNs, combined with the complexity of high-dimensional data, poses significant resource challenges in applying IDS. Specifically, the constrained computational, memory, and energy resources of WSNs make it difficult to efficiently apply classification algorithms. To overcome these limitations, integrating a suitable DR technique can indeed be a promising solution.

In our recent work [11], we demonstrated that the (ARF+HAT) ensemble outperformed other ensemble classifiers on WSN datasets, though this approach was applied without the use of DR techniques. Given their strong performance in dynamic environments, these algorithms handle diverse data types effectively and are robust to shifts in data distribution over time. However, due to the high-dimensional nature of data used in OL, applying DR becomes essential, not only to enhance model performance but also to reduce computational overhead. In this study, we aim to enhance the performance of the ARF+HAT ensemble by applying an appropriate DR technique.

Data-independent DR techniques, primarily based on random projections, are well-suited to the streaming and evolving nature of WSN data because they generate projection matrices (or functions) and transform data into a low-dimensional space independently of the input data. Unlike data-dependent methods, such as Principal Component Analysis (PCA), they do not require storing or re-estimating a covariance matrix, making them more scalable and less sensitive to new or unseen instances.

Definition 1 (Data Streams). *Whether online or incremental, refers to an infinite sequence of multidimensional, sporadic, and transient instances that become available over time. Let* $S = \{X_1, X_2, \ldots, X_N, \ldots\}$ *represent the stream, where each instance* X_i *is described by a features or attributes, expressed as* $X_i = (x_i^1, \ldots, x_i^a)$*, and* N *is the number of instances processed so far in the stream.*

Definition 2 (Dimensionality Reduction (DR)). *Given an instance* X_i *from the stream, the goal of DR is to determine a transformation function (or map)* $A : \mathbb{R}^a \rightarrow \mathbb{R}^m$*, with* $m \ll a$*, to be applied as follows:*

$$Y_i = A(X_i), \tag{1}$$

where $Y_i = (y_i^1, \ldots, y_i^m)$ *represents the reduced-dimensional version of* X_i*.*

In the following subsections, we describe the specific data-independent DR methods selected for this study and their suitability for WSNs' unique constraints.

3.1 Random Projection

As a prominent method in DR, Random Projection (RP) has been applied extensively alongside mining algorithms to address a range of complex challenges [15]. RP leverages the Johnson-Lindenstrauss (JL) Lemma [16], which states that a set of N points in Euclidean space can be projected into a lower-dimensional space of $\mathcal{O}(\log(N/\epsilon^2))$ dimensions, preserving pairwise distances within a multiplicative factor of $1 \pm \epsilon$ [17].

Consider $\epsilon \in [0, 1]$, and a set $S = \{X_1, ..., X_N\} \subset \mathbb{R}^a$. For a dimension $m \geq \log(N/\epsilon^2)$, we can define a linear map $A : \mathbb{R}^a \rightarrow \mathbb{R}^m$ such that for every pair $X_i, X_j \in S$, we have:

$$(1 - \epsilon)\|X_i - X_j\|_2^2 \leq \|AX_i - AX_j\|_2^2 \leq (1 + \epsilon)\|X_i - X_j\|_2^2, \tag{2}$$

where A represents a random matrix, typically generated using a Gaussian distribution. Thus, RP provides a computationally efficient and straightforward method to reduce the dimensionality of input data while roughly maintaining pairwise distances between instances.

- **Gaussian Random Projection (GRP)**: This method employs a projection matrix whose entries are independently drawn from a Gaussian distribution, having independent and identically distributed (i.i.d) values with a mean of zero and variance of one:

  $$A_{i,j} \sim \mathcal{N}(0, 1)$$

  This construction ensures that each entry in A contributes randomly but consistently across dimensions, preserving distances between points with high probability when the data is projected to a lower-dimensional space.
- **Random Bernoulli Projection (RBP)**: Unlike GRP, this approach uses a Bernoulli-distributed matrix, where each entry is randomly sampled with equal probability:

  $$A_{i,j} \in \left\{ \frac{1}{\sqrt{a}}, -\frac{1}{\sqrt{a}} \right\}$$

  where a is the original feature dimension. This construction provides a computational advantage, as it reduces floating-point calculations, making it less computationally intensive than Gaussian projection.

3.2 Hashing Trick

A widely recognized technique is the Hashing Trick (HT), also known as the feature hashing [18], is a reliable and efficient technique for projecting sparse vectors into lower-dimensional space using a hash function. HT maps an input vector $\mathbf{X} \in \mathbb{R}^a$ to a compressed vector $\mathbf{Z} \in \mathbb{R}^m$ (where $m \ll a$) by applying a hash function h and a sign function ξ:

$$Z_j = \sum_{i:h(i)=j} \xi(i)\, X_i$$

where,

- $h : \{1, \ldots, a\} \to \{1, \ldots, m\}$ assigns each feature to one of m dimensions.
- $\xi(i) \in \{-1, +1\}$ is a sign function that reduces collisions by assigning a random sign to each feature.

The fundamental concept of the HT is shown in Fig. 1. An important aspect of HT is that model quality can vary with the size of the hash table: generally, larger hash table sizes yield better models. However, an optimal hash table size can be selected to achieve near-perfect performance without significantly increasing the output dimension size. HT is valued for its simplicity, speed, and ability to preserve sparsity, making it very memory-efficient as it produces a fixed-size

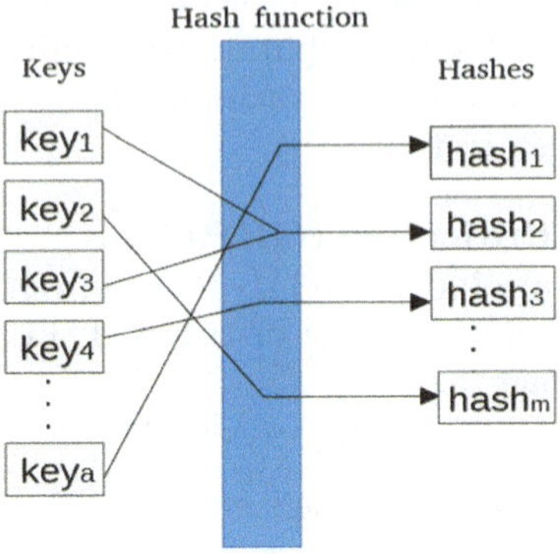

Fig. 1. Hashing trick.

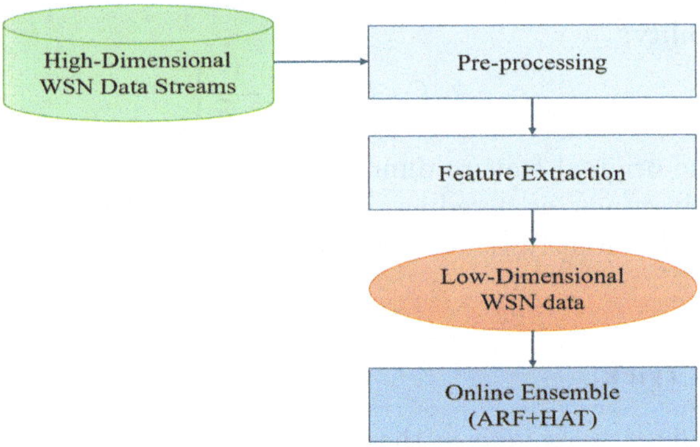

Fig. 2. Decomposition of the ARF+HAT Online Ensemble Scheme.

output vector. These properties make HT particularly suitable for online learning on data streams in memory-constrained environments.

Figure 2 presents the main flow of our proposed approach, combining online ensemble (ARF+HAT) with various DR techniques to improve detection performance on WSN data streams. To cope with drifts, the ensemble (ARF+HAT) methods are coupled with the change detector and the estimator ADaptive WINdowing (ADWIN) [19] applying default parameters: (i) the warning threshold $\delta_w = 0.00001$, and (ii) the drift threshold $\delta_d = 0.0001$. For each algorithm, we used the default settings. The maximum number of samples for the algorithms is also determined with regard to the total size of the given dataset. Projections were created for eight target dimensions $\{2, 4, 8, 10, 12, 14, 16\}$, balancing processing time and memory usage while retaining essential features for accurate classification of WSN data streams.

4 Experiments and Results

4.1 Experimental Environment

The hardware used in this paper consisted of a computer equipped with an AMD Ryzen™ 9 6900HS CPU with Radeon™ Graphics (3.30 GHz), 16 GB of RAM, and a NVIDIA® GeForce RTX™ 3060 Laptop GPU with 8 GB of memory. The software environment included Windows 11, utilizing Python, and extending the River framework[1] [20], alongside libraries such as `math`, `matplotlib`, and others.

4.2 Dataset Characteristics

In this work, we utilized the dataset created by Almomani et al. [21], which was specifically designed for DoS attack detection in WSNs. The network was simulated using Network Simulator 2 (NS-2), with the Low-Energy Adaptive Clustering Hierarchy (LEACH) protocol, to collect data across various attack scenarios. The dataset contains 23 sensor state features, though only 19 were available in the CSV file, and 374,661 instances, representing normal operation and four distinct types of DoS attacks: Blackhole, Grayhole, Flooding, and Scheduling, making it appropriate for evaluating IDSs in WSN environments. Label encoding was employed to convert categorical label values into numeric ones, ensuring they did not influence the algorithm's performance. Table 1 provides an overview of the statistical attributes of the WSN-DS dataset.

Table 1. Distribution of WSN-DS dataset

Samples Type	Quantity	Percentage (%)
Normal	340,066	90.77
Grayhole	14,596	3.90
Blackhole	10,049	2.68
Scheduling	6,638	1.77
Flooding	3,312	0.88
Total Traces	**374,661**	**100**

Figure 3 illustrates the 2-dimensional projection of the WSN-DS dataset using UMAP, t-SNE, and PCA, with colors representing the corresponding labels. These methods reduce the high-dimensional feature space to two dimensions, allowing us to visualize patterns and relationships among the different categories within the dataset.

- **Principal Components Analysis (PCA)** is one of the most widely-used and straightforward unsupervised techniques for DR by identifying a lower-dimensional space that minimizes the sum of squared distances between the

[1] https://riverml.xyz/.

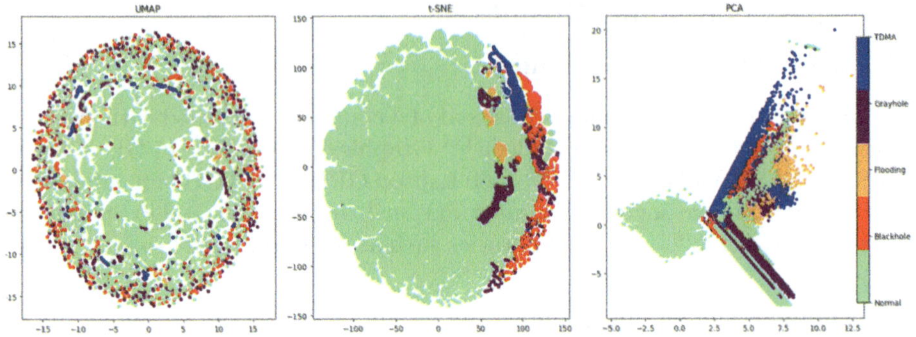

Fig. 3. Visualization of WSN-DS dataset using UMAP, t-SNE and PCA.

original data points and their projections, ideally approaching zero, while capturing as much variance as possible in the first components. Mathematically, PCA seeks to find a linear transformation consisting of a few orthogonal linear combinations (eigenvectors or principal components) from the original data that optimizes a specific cost function.

- **t-Distributed Stochastic Neighbor Embedding (t-SNE)** [22] is one of the most distinguished DR techniques in the current state-of-the-art. This graph-based non-linear approach intended to visualize high-dimensional data in a lower-dimensional space (usually 2 or 3 dimensions) by ensuring that similar instances in the original high-dimensional space remain close to each other in the reduced space. In t-SNE, the perplexity parameter, comparable to the number of neighbors, controls the size of the local neighborhood for each node in the graph. As a result, t-SNE emphasizes local structure at the expense of preserving the global data layout.
- **Uniform Manifold Approximation and Projection (UMAP)** [23] is a recently developed manifold technique, akin to t-SNE, that has gained significant attention and is based on a solid mathematical framework rooted in Riemannian geometry. UMAP constructs simplicial complexes from open balls around instances to capture the topological structure of the original space, then optimizes the representation in a lower-dimensional space by minimizing cross-entropy between fuzzy representations. As a non-linear DR technique, UMAP preserves both global and local structure in the data, making it faster than t-SNE and useful not only for visualization but also as a general DR technique for data mining algorithms.

From these visualizations, it becomes clear that the WSN-DS dataset contains complex, non-linear relationships between normal instances and attack types, highlighting the challenges of differentiating between the various attack types in the dataset. In Fig. 3, we observe that t-SNE offers the clearest separation of different categories, showing its effectiveness in distinguishing between attack types and normal instances (5 classes), outperforming UMAP and PCA.

UMAP presents a more spread-out view with greater overlap between attack types and normal instances compared to t-SNE. The overlap in UMAP indicates some challenges in class separation, which could affect later classification tasks, especially for distance-based algorithms, as global distances and density are sacrificed to preserve both local and global structure. This overlap suggests that some attacks share similar features, making it harder to fully differentiate them. On the other hand, PCA, as a linear transformation, struggles to clearly separate instances, preventing the formation of distinct clusters. However, it still serves as a useful baseline for understanding the variance within the dataset. The effectiveness of UMAP and t-SNE in revealing clusters suggests that there are distinct patterns in the feature space, although the overlaps indicate that some attacks share similar features.

4.3 Performance Metrics and Evaluation

Known as interleaved test-then-train evaluation, prequential evaluation [24] is a commonly applied technique in data stream analysis. In this method, instances are first used to test the current model before being employed to update it. In other words, each instance serves two purposes. To evaluate the performance of our proposed methodology, we focus on using six closely related evaluation criteria: Accuracy, Precision, Recall, F1-Score, Running Time and Memory. The first four metrics are mathematically defined in Table 2, based on the following terms:

- True Positives (TP): Instances where attacks are correctly classified as attacks.
- True Negatives (TN): Instances where normal cases are correctly classified as normal.
- False Positives (FP): Instances where normal cases are incorrectly classified as attacks.
- False Negatives (FN): Instances where attacks are incorrectly classified as normal cases.

Table 2. Calculation equations for performance metrics

Performance metrics	Mathematical expression
Accuracy (Acc)	$\frac{TP+TN}{TP+FN+TN+FP}$
Recall or Detection rate (DR)	$\frac{TP}{TP+FN}$
Precision (Pr)	$\frac{TP}{TP+FP}$
F1-Score (F1)	$2 \times \frac{\text{Precision} \times \text{Recall}}{\text{Precision}+\text{Recall}}$

- Running Time (RT): A good classification algorithm minimizes the time it takes to process instances as they arrive. This time includes any preprocessing (e.g., dimensionality reduction), learning, and prediction operations performed on new data.
- Memory: Measured in megabytes (MB), memory usage by an algorithm consists of (i) storage for the current model and/or (ii) memory allocated for storing running statistics that facilitate the incremental handling of data streams.

4.4 Results and Findings

A series of detailed experiments was conducted to evaluate the performance of our proposed approaches. Specifically, the experiments assessed the three decomposition techniques, GRP, HT, and RBP, across six metrics: Acc, PR, DR, F1, memory, and RT. These results were compared to a baseline using a heterogeneous ensemble (ARF+HAT), which used the full dataset without DR as a reference. Each metric was assessed across eight output dimensionalities [2,4,6,8,10,12,14,16] to explore trade-offs in classifier performance, computational efficiency, and resource consumption. Figure 4 presents the detailed results, illustrating which methods perform best at specific dimensions and how trade-offs shift with increasing dimensionality.

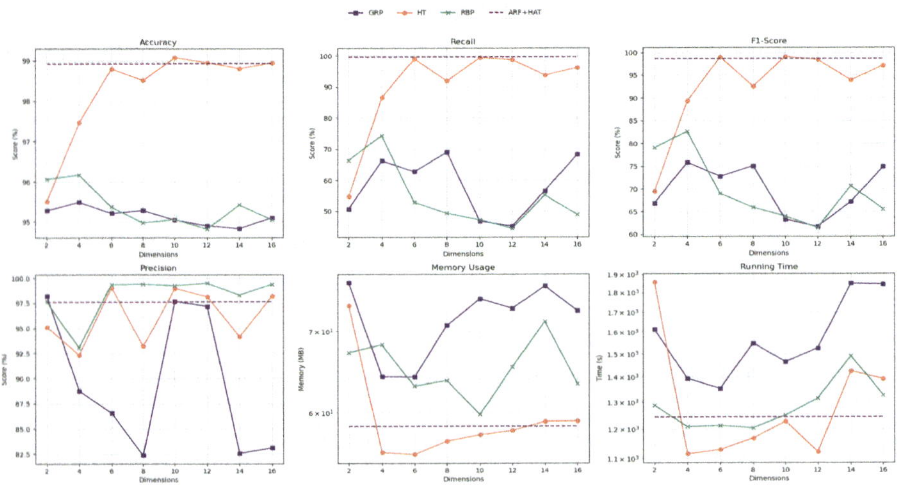

Fig. 4. Impact of GRP, HT, and RBP dimensionality reduction methods compared to the ARF+HAT baseline across varying output dimensions.

The HT method consistently excels across key metrics, including ACC, DR, F1, memory, and RT, making it the most balanced choice across all tested dimensions. It achieves scores close to the ARF+HAT baseline, especially in dimensions

8 and 10, while maintaining low memory usage and stable runtime compared to GRP and RBP, particularly in the lower and mid-range dimensions (4 to 10). This consistent performance underscores HT's reliability across diverse dimensional settings, making it well-suited for resource-constrained WSNs by balancing detection accuracy with efficient resource usage. While RBP excels in PR, making it ideal for applications needing minimal false positives, it may not consistently capture all relevant instances. Its performance in ACC, DR, and F1 is more variable, and its moderate memory and runtime suggest a reasonable alternative to HT when PR is prioritized. However, its less stable performance across other metrics and varying predictability make it a less reliable choice for general applications, where balanced ACC and efficiency are needed. GRP generally lags behind HT and RBP in terms of memory consumption and runtime, particularly at higher dimensions (e.g., dimensions 12, 14, and 16), making it less suited for resource-constrained settings in WSNs. It exhibits significant variability across metrics and consistently falls short in Acc, DR, F1 and PR compared to HT and the baseline. Additionally, GRP's high memory usage and extended runtimes, especially at higher dimensions, may limit its practicality in memory-constrained or time-sensitive environments.

Table 3. Optimal Trade-Off Points for Each Method at Selected Dimensions

Metric	ARF+HAT (All Dimensions)	HT (10)	GRP (4)	RBP (6)
Acc (%)	98.92	99.08	95.50	95.37
Pr (%)	97.62	99.00	88.81	99.36
DR (%)	99.62	99.38	66.29	52.83
F1 (%)	98.61	99.19	75.91	68.98
Memory (MB)	58.47	57.53	64.22	63.08
RT (s)	1244.23	1227.76	1393.71	1212.69

Table 3 presents the optimal dimension for each method, highlighting the best-performing configurations. Through extensive analysis across multiple dimensions, dimension 10 was identified as the optimal setting for the specific WSN-DS dataset. At this dimension, the HT method demonstrated an excellent balance between performance, processing time, and memory usage. Overall, HT proves to be the most efficient and reliable detection method for general WSNs real-time applications, while RBP may be preferable for tasks requiring high precision, whereas GRP is less suitable due to its resource-intensive nature and variability across metrics.

5 Conclusion

The primary challenge in developing a robust IDS for WSNs lies in balancing system longevity and detection reliability within resource constraints. Our research introduced a lightweight IDS that integrates an online ARF+HAT ensemble with dimensionality reduction techniques to enhance detection performance. Among the DR methods tested, the HT demonstrated balanced performance across metrics, making it the optimal choice for robust classification in resource-constrained WSNs. As the first to apply DR with an ARF+HAT ensemble for WSN anomaly detection, this study establishes a novel, scalable approach for real-time security. Future research could investigate adaptive DR to further improve IDS adaptability in dynamic WSN environments.

References

1. Huang, T., Yu, F.R., Zhang, C., Liu, J., Zhang, J., Liu, Y.: A survey on large-scale software defined networking (SDN) testbeds: approaches and challenges. IEEE Commun. Surv. Tutorials **19**(2), 891–917 (2016)
2. Ramson, S.J., Moni, D.J.: Applications of wireless sensor networks-a survey. In: 2017 International Conference on Innovations in Electrical, Electronics, Instrumentation and Media Technology (ICEEIMT), pp. 325–329. IEEE (2017)
3. Otoum, S., Ahmed, M., Mouftah, H.T.: Sensor medium access control (SMAC)-based epilepsy patients monitoring system. In: 2015 IEEE 28th Canadian Conference on Electrical and Computer Engineering (CCECE), pp. 1109–1114. IEEE (2015)
4. Ifzarne, S., Hafidi, I., Idrissi, N.: Homomorphic encryption for compressed sensing in wireless sensor networks. In: Proceedings of the 3rd International Conference on Smart City Applications, pp. 1–6 (2018)
5. Ifzarne, S., Hafidi, I., Idrissi, N.: Compressive sensing and Paillier cryptosystem based secure data collection in WSN. J. Ambient Intell. Humaniz. Comput. **14**(5), 6243–6250 (2023)
6. Ifzarne, S., Hafidi, I., Idrissi, N.: Secure data collection for wireless sensor network. In: Emerging Trends in ICT for Sustainable Development: the Proceedings of NICE2020 International Conference, pp. 241–248. Springer (2021)
7. Ifzarne, S., Hafidi, I., Idrissi, N.: A novel secure data aggregation scheme based on semi-homomorphic encryption in WSNs. J. Commun. **16**(8), 323–330 (2021)
8. Tabbaa, H., Hafidi, I.: An online model for detecting attacks in wireless sensor networks. In: International Conference of Machine Learning and Computer Science Applications, pp. 271–282. Springer (2022)
9. Ghosal, A., Halder, S.: A survey on energy efficient intrusion detection in wireless sensor networks. J. Ambient Intell. Smart Environ. **9**(2), 239–261 (2017)
10. Alom, M.Z., Taha, T.M.: Network intrusion detection for cyber security using unsupervised deep learning approaches. In: 2017 IEEE National Aerospace and Electronics Conference (NAECON), pp. 63–69. IEEE (2017)
11. Tabbaa, H., Ifzarne, S., Hafidi, I.: An online ensemble learning model for detecting attacks in wireless sensor networks. Comput. Inf. **42**, 1013–1036 (2023)
12. Ifzarne, S., Tabbaa, H., Hafidi, I., Lamghari, N.: Anomaly detection using machine learning techniques in wireless sensor networks. J. Phys. Conf. Ser. **1743**, 012021 (2021)
13. Martindale, N., Ismail, M., Talbert, D.A.: Ensemble-based online machine learning algorithms for network intrusion detection systems using streaming data. Information **11**(6), 315 (2020)

14. Ding, Z., Wang, H., Fei, M., Du, D.: A novel distributed online anomaly detection method in resource-constrained wireless sensor networks. Int. J. Distrib. Sens. Netw. **11**(10), 146189 (2015)
15. Vempala, S.S.: The Random Projection Method, vol. 65. American Mathematical Soc., USA (2005)
16. Johnson, W.B.: Extensions of Lipshitz mapping into Hilbert space. In: Conference Modern Analysis and Probability, pp. 189–206 (1984)
17. Johnson, W.B., Lindenstrauss, J., Schechtman, G.: Extensions of Lipschitz maps into Banach spaces. Israel J. Math. **54**(2), 129–138 (1986)
18. Weinberger, K., Dasgupta, A., Langford, J., Smola, A., Attenberg, J.: Feature hashing for large scale multitask learning. In: Proceedings of the 26th Annual International Conference on Machine Learning, pp. 1113–1120 (2009)
19. Bifet, A., Gavalda, R.: Learning from time-changing data with adaptive windowing. In: Proceedings of the 2007 SIAM International Conference on Data Mining, pp. 443–448. SIAM (2007)
20. Montiel, J., et al.: River: machine learning for streaming data in Python (2021)
21. Almomani, I., Al-Kasasbeh, B., Al-Akhras, M., et al.: WSN-DS: a dataset for intrusion detection systems in wireless sensor networks. J. Sens. **2016**(1), 4731953 (2016)
22. Maaten, L., Hinton, G.: Visualizing data using t-SNE. J. Mach. Learn. Res. **9**(11), 2579–2605 (2008)
23. McInnes, L., Healy, J., Melville, J.: UMAP: uniform manifold approximation and projection for dimension reduction. arXiv preprint arXiv:1802.03426 (2018)
24. Dawid, A.P.: Present position and potential developments: some personal views statistical theory the prequential approach. J. Roy. Stat. Soc. Ser. A Gen. **147**(2), 278–290 (1984)

Optimizing Coffee Waste Collection: A Discrete Event Simulation Model Based on Consumption Patterns

Manar Tayouga[1](✉), Mohamed Ouzineb[2], and Soumaya Yacout[1]

[1] MAGI, Ecole Polytechnique de Montreal, Montreal, Canada
[2] The National Institute of Statistics and Applied Economics, Rabat, Morocco
soumaya.yacout@polymtl.ca

Abstract. In recent years, global coffee consumption has soared, surpassing 10.7 million tons in 2022/23. This trend has led to an increase in spent coffee grounds (SCG), with each kilogram of soluble coffee yielding nearly 2 kg of wet waste. Much of this waste ends up in landfills, where it emits methane, a greenhouse gas that is 25 times more potent than carbon dioxide. This environmental challenge requires the development of an effective waste management system specifically designed for SCG.

This paper investigates a method to optimize coffee waste collection in urban areas, using a combined approach that integrates a vehicle routing problem (CVRP) with a discrete event simulation model (DES). The CVRP model is used to find the best collection routes minimizing the travel distance while respecting vehicle capacity and operating restrictions. The model is informed by data such as the locations of coffee shops and treatment centers, waste generation estimates based on consumption trends, and relevant demographic information.

The DES component simulates the collection process in various scenarios, offering performance insights when waste levels or operating conditions fluctuate. By combining these methods, this study provides a framework that serves as a scalable solution that can be adapted to other urban settings, contributing to the broader goal of sustainable urban waste management.

Keywords: Coffee waste collection optimization · Capacitated Vehicle Routing Problem (CVRP) · Discrete event simulation

1 Introduction

Coffee has become an integral part of daily routines around the world, with global consumption reaching unprecedented levels. However, the environmental impact of this trend is often overlooked. A major environmental concern associated with coffee consumption is the disposal of spent coffee grounds (SCG), a byproduct generated in substantial quantities. Improper management of SCGs, particularly their disposal in landfills, contributes to harmful greenhouse gas emissions, such

A. Laghrib and A. Ghazdali (Eds.): ICMDS 2024, LNNS 1466, pp. 226–236, 2025.
https://doi.org/10.1007/978-3-031-94802-2_22

as methane, exacerbating environmental concerns [1]. This challenge is especially pronounced in urban areas, where coffee shops are densely concentrated, generating large but unevenly distributed waste. Developing efficient collection systems for SCG presents both logistical and environmental challenges. Unlike conventional waste streams, SCG production can vary greatly due to differences in local consumption patterns, demographic factors, and operational conditions. Addressing these challenges requires innovative solutions that combine advanced optimization techniques with practical tools for dynamic analysis.

Over the past decade, research in waste management has progressed significantly, driven by technological advancements and a growing awareness of environmental issues. Early studies [2] introduced multi-agent Internet of Things (IoT) frameworks for smart waste monitoring and collection, achieving up to 93% efficiency in vehicle routing. These innovations underscored the importance of real-time data integration and active citizen participation in improving the efficiency of waste collection systems. The discourse has since broadened to encompass circular economy principles. For example, reverse logistics strategies [3] in coffee capsule recycling have placed greater emphasis on resource recovery and active consumer participation. Industry 4.0 technologies [4] have further enhanced municipal waste routing optimization and environmental sustainability, using system dynamics and network theory to model efficient operations. Specific efforts have been dedicated to the management of spent coffee grounds (SCG). Research has characterized SCGs as a valuable bio-resource [5,6], identifying applications in energy generation and the food industry [7]. These findings support broader initiatives aimed at extending the lifecycle of coffee-related products and minimizing environmental impacts. Additionally, studies on optimizing waste-to-energy plant locations [8] and developing robust frameworks for waste bin placement under uncertain conditions [9] have underscored the critical role of strategic planning in the design of efficient waste collection systems.

Efforts to integrate spatial and technological considerations into waste management have also gained traction. For instance, frameworks for wastewater treatment in eco-industrial parks [10] have combined transport network optimization with resource recovery to enhance system efficiency. Meanwhile, bibliometric analyses have underscored the growing role of machine learning and artificial intelligence in advancing waste management systems [11], demonstrating their potential to improve decision-making and operational performance.

Despite notable advancements in waste management optimization, the use of discrete event simulation (DES) models for SCG collection remains underexplored. These models offer a unique capability to capture the real-world dynamics of waste collection, enabling the optimization of routes and the maximization of vehicle utilization based on fluctuating consumption patterns.

This study proposes a hybrid framework to optimize the collection of SCG in urban settings by integrating a Capacitated Vehicle Routing Problem (CVRP) model with Discrete Event Simulation (DES). The CVRP model designs optimal collection routes by minimizing travel distances while accounting for vehicle capacity and operational constraints. The DES model complements this approach by simulating real-world scenarios, enabling planners to evaluate and refine collection strategies in response to varying inputs and conditions.

By utilizing estimated data on SCG generation and geospatial information, this framework seeks to deliver actionable insights for designing sustainable and scalable waste management systems. This paper focuses on demonstrating the potential of integrating optimization and simulation to enhance operational efficiency and reduce environmental impacts.

The remainder of this paper is organized as follows: Sect. 2 details the methodology, covering the CVRP and DES models. Section 3 explains the data sources and experimental setup. Section 4 presents the results and discussion, while Sect. 5 concludes with key insights and potential directions for future research.

2 Methodology

This section presents the methodological framework used to optimize the collection of spent coffee grounds (SCG) in urban environments. The proposed approach combines a Capacitated Vehicle Routing Problem (CVRP) model with Discrete Event Simulation (DES) to account for the dynamic nature of waste collection driven by real-world consumption patterns.

2.1 Framework Overview

The hybrid framework consists of two primary components:

- **Capacitated Vehicle Routing Problem (CVRP) Model**: Used to design efficient collection routes while minimizing operational costs and adhering to vehicle capacity constraints.
- **Discrete Event Simulation (DES)**: Simulates the collection process under variable conditions to provide insights into system performance and refine the CVRP-generated routes.

The CVRP model produces an optimized set of initial routes, while the DES model evaluates and refines these routes in response to dynamic, real-world factors. This integration allows for iterative route adjustments, ensuring a balance between operational efficiency and consistent service quality.

2.2 Capacitated Vehicle Routing Problem (CVRP) Model

The objective of the CVRP model is to determine optimal routes for a fleet of vehicles tasked with collecting SCG from multiple locations. The model seeks to minimize the total travel distance and associated costs while ensuring compliance with vehicle capacity limits and operational constraints.

Notation

Let:

- N: Set of all locations, where $N = \{0, 1, 2, \ldots, n\}$ with 0 representing the depot.
- V: Set of vehicles, where each vehicle $v \in V$ has a capacity Q_v.
- d_{ij}: Distance between locations i and j.
- q_i: Quantity of SCG at location i.
- x_{ij}^v: Binary decision variable, equal to 1 if vehicle v travels directly from location i to location j, and 0 otherwise.
- y_i^v: Load of vehicle v after visiting location i.

Objective Function

The objective is to minimize the total distance traveled by all vehicles:

$$\min \sum_{v \in V} \sum_{i \in N} \sum_{j \in N} d_{ij} \cdot x_{ij}^v$$

Constraints

1. *Route Completion*: Each location is visited exactly once by a single vehicle.

$$\sum_{v \in V} \sum_{j \in N} x_{ij}^v = 1, \quad \forall i \in N \setminus \{0\}$$

2. *Flow Conservation*: Ensures continuity in the routes for each vehicle.

$$\sum_{j \in N} x_{ij}^v = \sum_{j \in N} x_{ji}^v, \quad \forall v \in V,\ i \in N$$

3. *Capacity Constraints*: The quantity collected by each vehicle must not exceed its capacity.

$$y_i^v + q_j \cdot x_{ij}^v \leq y_j^v, \quad \forall v \in V,\ i, j \in N,\ i \neq j$$

$$y_i^v \leq Q_v, \quad \forall v \in V,\ i \in N$$

4. *Depot Constraints*: Vehicles start and end their routes at the depot.

$$\sum_{j \in N} x_{0j}^v = 1, \quad \forall v \in V$$

$$\sum_{i \in N} x_{i0}^v = 1, \quad \forall v \in V$$

5. *Subtour Elimination* (Optional for larger problems): Prevents small loops that do not include the depot.

$$y_i^v - y_j^v + Q_v \cdot x_{ij}^v \leq Q_v - q_j, \quad \forall i, j \in N,\ v \in V,\ i \neq j$$

Explanation

The *Objective Function* minimizes total travel distance, optimizing cost and efficiency. The *Route Completion* constraint ensures each location is visited once, while *Flow Conservation* maintains continuity in each route. *Capacity Constraints* limit SCG collection to vehicle capacity, and *Depot Constraints* require each route to start and end at the depot. This formulation establishes optimized SCG collection routes, serving as a baseline for refinement in the DES model.

2.3 Discrete Event Simulation (DES) Model

The DES model is designed to simulate the collection process in real time, capturing daily variations in SCG generation. This model supports:

- **Scenario Testing**: Evaluation of different operational strategies under varying conditions, such as peak collection times or route deviations.
- **Vehicle Utilization Optimization**: Analysis of vehicle loading and unloading times, idle times, and overall utilization.
- **Adaptability to Consumption Patterns**: Incorporation of fluctuations in SCG quantities based on coffee consumption trends, enabling a dynamic response during high-demand periods.

The DES model is implemented using AnyLogic, which enables event-driven simulation to capture the temporal dynamics of the waste collection process.

2.4 Model Integration

The hybrid approach is implemented in two stages:

1. **Initial CVRP Optimization**: The CVRP model generates baseline routes for waste collection by minimizing travel distance and adhering to operational constraints.
2. **DES-Based Route Refinement**: The DES model simulates the collection process, providing feedback on route feasibility and suggesting adjustments based on dynamic, simulated data.

The iterative feedback loop between the CVRP and DES models enables continuous refinement, accommodating fluctuations in SCG quantities and dynamically optimizing vehicle allocation. This hybrid approach is designed to improve both route efficiency and adaptability to real-world conditions.

3 Data Sources and Experimental Setup

This section describes the data sources and experimental setup used to validate the proposed hybrid model for SCG collection optimization. In the absence of real-time data, estimations were employed, drawing from existing literature, regional consumption trends, and informed assumptions. The experimental setup integrates the CVRP and DES models to evaluate the proposed collection routes and assess their effectiveness in minimizing travel distances and improving overall operational efficiency.

3.1 Data Sources and Estimation Methodology

The following data elements were collected and estimated to simulate a realistic SCG collection scenario:

- **Geolocation Data of Collection Points**: Coffee shop locations were approximated using publicly available data from platforms such as Google Maps. Latitude and longitude coordinates for each coffee shop were extracted to create a spatial map of collection points.
- **SCG Quantity Estimation**: The quantity of SCG generated per coffee shop was estimated based on data from coffee consumption studies. These estimates were refined using factors such as Morocco's coffee culture (particularly *out-of-home* consumption), the population of the study area, shop popularity, and local consumption patterns. Assumptions regarding SCG generation per cup of coffee were derived from relevant research findings.
- **Distance Matrix**: A matrix of travel distances between all collection points and the depot was generated. Distances and travel times were computed using the Google Maps API to ensure realistic estimates of route lengths and durations.

While these data points are estimated, they provide a robust framework for testing the hybrid CVRP and DES model, with the primary objective of minimizing travel distance.

3.2 Experimental Setup

The hybrid model was implemented in two stages, using IBM CPLEX Optimization Studio for the CVRP and AnyLogic for the DES simulations. Key components of the experimental setup are outlined as follows:

- **Initial Route Optimization with CVRP**: The CVRP model was first used to determine the shortest possible routes, minimizing the total travel distance while ensuring that each vehicle adhered to its capacity constraints. This optimization provided baseline routes for SCG collection.
- **Discrete Event Simulation for Route Refinement**: The DES model, implemented in AnyLogic, simulated the collection process under dynamic, real-world conditions. It adjusted the baseline CVRP routes based on variations in SCG quantities and real-time collection dynamics, iteratively improving route efficiency.

- **Iterative Feedback Loop**: Following each simulation run, the DES model generated feedback on route efficiency, which was fed back into the CVRP model to refine routes further. This iterative loop continued until the routes achieved stable, minimized travel distances within acceptable operational limits.

3.3 Performance Metrics

The primary performance metric for this study was the **total travel distance**, representing the effectiveness of the hybrid model in achieving its objective. Secondary metrics included:

- **Vehicle Utilization Rate**: The percentage of vehicle capacity used, indicating how efficiently each vehicle operated within its load constraints.
- **Average Collection Time**: The total time spent per route, including loading and transit times, providing insights into the operational feasibility of the collection process.
- **Route Stability**: Measured by the number of iterations required for route convergence, indicating the model's stability in reaching optimized routes.

The experimental setup enabled testing under various simulated conditions to evaluate the hybrid model's adaptability to changes in SCG generation patterns and its effectiveness in minimizing travel distance.

4 Results and Discussion

This section presents the results of the hybrid CVRP and DES model applied to optimize the collection of spent coffee grounds (SCG) in Morocco. The model's performance was evaluated using key metrics, including total travel distance, vehicle utilization, cost efficiency, and adaptability to variations in waste generation periods.

4.1 Performance Metrics

The primary performance metric for this study was the **total travel distance**, representing the effectiveness of the hybrid model in achieving its objective. Secondary metrics included (Table 1):

This experimental setup enabled testing under various simulated conditions, demonstrating the hybrid model's adaptability to different SCG generation patterns.

Table 1. Performance Metrics for Waste Collection Optimization

Metric	High Waste Period	Low Waste Period	Description
Total Distance (km)	780.4	410.3	Measures travel distance
Vehicle Utilization (%)	92	86	Utilization rate of vehicle capacity
Average Collection Time (hrs)	9.2	6.5	Time per route including idle times
Route Stability (iterations)	3	2	Number of iterations to convergence

4.2 Optimal Vehicle Count and Capacity

Vehicles with capacities ranging from 600 to 800 kg were used, operating within a 10-hour shift and maintaining an average speed of 35 km/h. Table 2 presents the optimal number of vehicles needed to meet demand efficiently during high and low waste periods.

Table 2. Optimal Number of Vehicles Needed

Vehicle Capacity	High Waste Period	Low Waste Period
600 kg	4 vehicles	2 vehicles
700 kg	4 vehicles	2 vehicles
800 kg	3 vehicles	2 vehicles

4.3 Cost Analysis for High and Low Waste Periods

The cost analysis is divided into two categories: high and low waste periods. The costs considered include 7 MAD per kilometer for fuel, 250 MAD per vehicle for maintenance, and 400 MAD per driver per shift. Additionally, idle time is estimated at 30 min per vehicle per route, resulting in an extra cost of 35 MAD.

Table 3. Cost Analysis for High Waste Period - Part 1

Scenario	Capacity (kg)	Vehicles	Distance (km)	Waste Collected (kg)
Scenario 1	600	3	780.4	1785
Scenario 2	700	3	700.5	1920
Scenario 3	800	4	845.6	2200

Table 4. Cost Analysis for High Waste Period - Part 2

Scenario	Total Cost (MAD)	Cost per kg (MAD/kg)
Scenario 1	7412.8	4.15
Scenario 2	6853.5	3.57
Scenario 3	8519.2	3.87

4.4 Key Insights from Cost Analysis

The results in Tables 3, 4, 5 and 6 highlight key findings:

- **High Waste Period**: Scenario 2 (700 kg capacity, 3 vehicles) achieved the lowest cost per kilogram (3.57 MAD/kg) while effectively balancing vehicle utilization and cost efficiency.

Table 5. Cost Analysis for Low Waste Period - Part 1

Scenario	Capacity (kg)	Vehicles	Distance (km)	Waste Collected (kg)
Scenario 1	600	2	410.3	950
Scenario 2	700	2	395.2	1050
Scenario 3	800	3	450.6	1200

Table 6. Cost Analysis for Low Waste Period - Part 2

Scenario	Total Cost (MAD)	Cost per kg (MAD/kg)
Scenario 1	4172.1	4.39
Scenario 2	4066.4	3.87
Scenario 3	5104.2	4.25

- **Low Waste Period**: During low waste periods, Scenario 2 (700 kg capacity, 2 vehicles) minimized costs, achieving a rate of 3.87 MAD/kg and demonstrating efficient alignment between vehicle capacity and demand.
- **Idle Time Impact**: The inclusion of estimated idle time led to a slight increase in overall costs, emphasizing the importance of minimizing delays to improve operational efficiency.

4.5 Conclusion on Cost-Efficient Configurations

The most cost-effective configurations based on this analysis are:

- **High Waste Period**: Scenario 2 with 700 kg capacity and 3 vehicles, achieving a cost efficiency of 3.57 MAD/kg.
- **Low Waste Period**: Scenario 2 with 700 kg capacity and 2 vehicles, achieving a cost efficiency of 3.87 MAD/kg.

This analysis supports scenario-based planning in waste collection, demonstrating how adjusting vehicle capacity and numbers to demand improves cost efficiency and resource utilization.
This cost analysis validates the effectiveness of the hybrid CVRP-DES model, supporting informed decision-making for operational efficiency and budget management.

5 Conclusion

This study introduced a hybrid model combining Capacitated Vehicle Routing Problem (CVRP) optimization with Discrete Event Simulation (DES) to enhance the efficiency of spent coffee grounds (SCG) collection in urban Morocco. The model minimized travel distance and operational costs by adjusting vehicle configurations in response to waste generation patterns. Results showed that, during high waste periods, the configuration with 700 kg capacity and three vehicles achieved optimal cost efficiency, while two vehicles were sufficient for low waste periods. Incorporating idle time provided a realistic cost assessment, emphasizing the need to reduce delays. Performance metrics indicated high vehicle utilization rates and route stability, confirming the model's robustness under variable demand. This approach highlights the utility of integrating optimization and simulation for sustainable waste management and presents a scalable framework applicable to other urban waste systems. Future work could enhance this model by including real-time data, traffic conditions, and environmental impact analysis to support dynamic and eco-friendly waste collection strategies.

Supplementary information

Not applicable

Acknowledgements

Not applicable

Declarations

Not applicable

Ethical approval declarations Not applicable

References

1. Duran-Aranguren, D.D., et al.: Scientometric overview of coffee by-products and their applications. Molecules **26**(24), 7605 (2021)
2. David Likotiko, N.D.E., Mwangoka, J.: Multi-agent based IoT smart waste monitoring and collection architecture. Int. J. Comput. Sci. Eng. Inf. Technol. **7**(5) (2017)
3. Paucar-Caceres, A., Toni Burrowes-Cromwell, T.: Consumers' values and behaviour in the Brazilian coffee-in-capsules market: promoting circular economy. Int. J. Prod. Res. **57**, 1–20 (2019)
4. Bányai, T., Tamás, P., Illés, B., Stankevičiūtė, Ž, Bányai, Á.: Optimization of municipal waste collection routing: impact of industry 4.0 technologies on environmental awareness and sustainability. Int. J. Environ. Res. Public Health **16**(4), 634 (2019)
5. O'Dwyer, E., Chen, K., Wang, H., Wang, A., Shah, N., Guo, M.: Optimisation of wastewater treatment strategies in eco-industrial parks: technology, location and transport. Chem. Eng. J. **381**, 122643 (2020)
6. Noman, A.A., Akter, U.H., Pranto, T.H., Haque, A.B.: Machine learning and artificial intelligence in circular economy: a bibliometric analysis and systematic literature review. An. Emerg. Technol. Comput. (AETiC) **6**(2) (2022)

7. Iriondo-DeHond, A., Iriondo-DeHond, M., Del Castillo, M.D.: Applications of compounds from coffee processing by-products. Biomolecules **10**(9), 1219 (2020)
8. Colantoni, A., et al.: Spent coffee ground characterization, pelletization test and emissions assessment in the combustion process. Sci. Rep. **4**(11), 5119 (2021)
9. Pluskal, J., et al.: Optimal location and operation of waste-to-energy plants when future waste composition is uncertain. Oper. Res. Int. J. **22**(5), 5765–5790 (2022)
10. Franca, A.S., Oliveira, L.S.: Potential uses of spent coffee grounds in the food industry. Foods **11**(14), 2064 (2022)
11. Rossit, D., Bard, J.: Solving the waste bin location problem with uncertain waste generation rate: a bi-objective robust optimization approach. Waste Manag. Res. **43**(3), 421–437 (2024)

Data Exploration by Unifying Clustering and Association Rule Mining

Youssef Fakir(✉), Salim Khalil, and Hamid Garmani

Department of Computer Sciences, Faculty of Sciences and Technics,
Sultan Moulay Slimane University, Beni Mellal, Morocco
info.dec07@yahoo.fr, h.garmani@usms.ma

Abstract. The extraction of association rules remains a crucial strategy in data analysis, particularly in the context of massive datasets. This method unveils complex relationships, correlations, and meaningful patterns within vast datasets, providing essential insights for decision-making and understanding behaviors. Our approach stands out with the use of clustering algorithms for intelligent data partitioning. This strategic choice establishes a robust foundation for efficient association rule extraction. By organizing the data specifically through clustering techniques before applying the extraction algorithm, we aim to optimize the relevance and significance of the discovered rules.

Keywords: Association rules · Apriori · FP-Growth · k-means · clustering

1 Introduction

In the era of data explosion, where datasets reach massive proportions, data mining has become a complex and crucial endeavor for organizations seeking meaningful insights. At the core of this quest lies association rule extraction, a fundamental data mining strategy for revealing subtle connections, unsuspected correlations, and essential patterns within vast and intricate datasets [1]. Our study focuses on an innovative approach, leveraging the K-means clustering algorithm for intelligent data partitioning [2]. This strategic approach aims to transcend the challenges posed by massive datasets by organizing information in a specific manner [3]. The ultimate goal is to maximize the relevance of extracted association rules, providing a deep and targeted understanding of dynamics within these extensive datasets [4,5].

Throughout this article, we delve into our approach, highlighting the benefits of intelligent data partitioning using the K-means algorithm in the context of data mining [6]. We demonstrate how this method establishes a robust foundation for efficient association rule extraction, paving the way for a more pertinent and meaningful exploration of massive datasets [7].

This study is structured to provide a comprehensive understanding of our innovative approach to association rule extraction in a distributed environment.

A. Laghrib and A. Ghazdali (Eds.): ICMDS 2024, LNNS 1466, pp. 237–248, 2025.
https://doi.org/10.1007/978-3-031-94802-2_23

In Sect. 2, we introduce the general context of massive data analysis and the importance of association rule extraction. Section 3 details our approach, highlighting the use of the K-means algorithm for intelligent data partitioning and the combination of association rule extraction algorithms such as Apriori and FP-Growth. We also present our methodology by describing the modeling steps used, while Sect. 4 examines the obtained results and offers detailed analyses. Finally, Sect. 5 concludes by emphasizing the importance of our integrated approach and future research prospects in this field. To further demonstrate the effectiveness and importance of clustering algorithms in data exploration and to highlight the valuable insights gained from the implementation of these algorithms, we will now examine recent research in this area, including related practical application [1,5,8,9].

2 State of the Art

The state of the art in the field of association rule mining highlights the increasing importance of using clustering techniques to enhance the efficiency of this process. Recent works have emphasized the positive impact of integrating the K-means algorithm into association rule mining, especially for analyzing large datasets. For example, Li et al. proposed a closed-loop hierarchical clustering approach for optimizing unit commitment and dispatch in micro grids [4]. Similarly, Guha et al. [10] presented a robust clustering algorithm for categorical attributes, called ROCK, which demonstrated significant performance in data segmentation.

In comparison with these recent works, the present article advocates for an innovative approach that synergistically combines clustering with association rule mining algorithms such as Apriori and FP-Growth. This approach aims to optimize the relevance and significance of discovered rules by preorganizing data using clustering techniques. By integrating these two aspects, this article offers a comprehensive methodology for association rule mining in a distributed environment, thus providing deeper insights into user behaviors and trends within massive datasets. Furthermore, the article underscores the importance of understanding the unique characteristics of association rule mining algorithms such as Apriori and FP-Growth and adapting the approach based on dataset specifics. This integrated approach highlights the effectiveness of data exploration through the combination of clustering and association rule mining techniques, offering promising prospects for more efficient and relevant data exploration practices.

3 Proposed Approach

Presently, databases pose a significant challenge due to their extensive size and distribution across various locations [11]. This has resulted in the extraction of association rules from these databases becoming an intricate and resource-intensive task marked by prolonged execution times and computational complexities [8]. Addressing these issues is imperative, and parallelism has emerged as a crucial solution for efficient association rule extraction [6,12].

Our exploration of sequential and parallel techniques revealed persistent challenges, notably the high volume of data scans and the detrimental impact of multiple synchronization and communication stages on algorithm performance [4,5]. Recognizing these obstacles, we identified the sequential partition algorithm as a potential remedy, requiring just two passes through the database [10]. To further alleviate this situation, our focus shifted toward enhancing the partition algorithm to streamline the process to a single database scan, with subsequent consideration for the development of a parallel version [2]. Additionally, optimizing algorithm performance demands intelligent base partitioning using clustering, particularly distributed clustering [3].

We closely examined sequential and parallel partitioning algorithms, recognizing them as promising avenues for achieving homogeneous database distribution across multiple sites [4,5]. As a proactive step, we propose a parallel clustering algorithm based on the k-prototype algorithm, which is deemed most effective for handling the complex nature of real-world data containing both numerical and categorical values [2]. The flowchart of the proposed model is shown in Fig. 1.

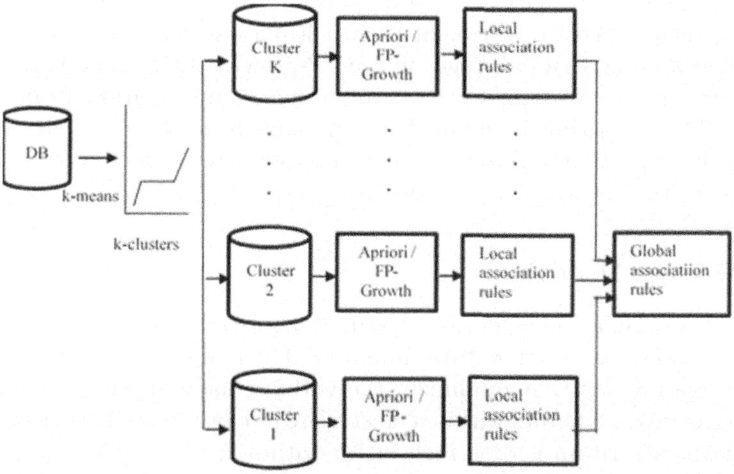

Fig. 1. Methodology approach.

3.1 FP-Growth Algorithm

The process of mining frequent itemsets in FP-Growth adopts a divide-and-conquer approach. Initially, FP-Growth condenses the database, transforming frequent itemsets into an FP-tree while preserving association information [13]. Association rules has been oriented towards two objectives:

- Determine the set of frequent itemsets that appear in the database with support greater than or identical to minsup (minimum support) where support is the proportion of transactions that contain the itemset, calculated using the formula:

$$Supp(X) = \frac{Number\ of\ transactions\ containing\ X}{Total\ number\ of\ transactions} \quad (1)$$

- Generate the set of associative rules, from these frequent itemsets, with a confidence measure greater than or identical to minconf. This measures the strength of the association between two itemsets. It is the conditional probability of finding Y in a transaction, given that X is present. The formula is:

$$Conf(X \Rightarrow Y) = \frac{Support(X \cup Y)}{Support(X)} \quad (2)$$

  where X and Y are the itemsets for which the confidence of the rule $X \Rightarrow Y$ (meaning "If X, then Y") is calculated.

Afterward, the compressed database is partitioned into a series of conditional databases, each linked to a frequent item. Ultimately, each database undergoes independent mining. Critical stages in the FP-growth algorithm involve building the FP-tree and subsequently extracting valuable information from this tree structure. The algorithm is detailed in our published paper [14]. It allows the discovery of frequent itemsets without generating candidate itemsets. The FP-Growth algorithm is described in Algorithm 1.

3.2 K-Means Algorithm

K-means is a popular unsupervised learning algorithm used to group data into k distinct clusters based on feature similarity. The main goal is to partition data into clusters in a way that minimizes the within-cluster variance, or in simpler terms, to ensure data points within a cluster are as close to each other as possible. The K-means algorithm is described in Algorithm 2.

3.3 A Priori Algorithm

The Apriori algorithm is widely used to mine frequent itemsets and generate association rules from transaction data. The Apriori algorithm consists of two main steps:

- **Step 1**: Frequent Itemset Generation: Identify all itemsets that meet the minimum support threshold.
- **Step 2**: Association Rule Generation: Generate rules from these frequent itemsets that meet the minimum confidence threshold.

Algorithm 1. FP-Growth Algorithm

Input: Database of transactions, minimum support threshold (minsup)
Output: Frequent itemsets, association rules

1. **Construction of the FP-tree:**
   - Traverse transactions, count the frequency of each unique item.
   - Remove infrequent items.
   - Sort the items based on frequency.
   - Construct an FP-tree.
2. **Construction of the Header Table:**
   - Create a header table to record the first occurrence of each item in the tree.
3. **Construction of Conditional Sets:**
   - For each item in the header table, extract conditional paths in the tree.
4. **Recurrence to Extract Frequent Sets:**
   - Repeat the process from Step 1 to extract frequent patterns from conditional sets.
5. **Generation of Association Rules:**
   - Generate all possible combinations of items in frequent patterns.
   - Split into antecedent and consequent patterns.
   - Calculate the confidence and filter the rules.
6. **Return the Final Set of Association Rules:**
   - The final output is a set of association rules meeting the specified criteria.

The Apriori algorithm is described in Algorithm 3.

3.4 Modeling with K-Means, Apriori and FP-Growth for Association Rule Extraction

The modeling process is as follows:

- **Step 1: Applying K-Means for Cluster Creation**: We begin our modeling by applying the K-Means algorithm to our original dataset [4]. The objective here is to group the data into separate clusters, each representing a homogeneous set of characteristics. After applying K-means, each data point is assigned to a specific cluster based on its characteristics, and creating a clear segmentation of our dataset [6].
- **Step 2: Creating Subsets of Data**: Once K-Means has created clusters, we divide the original dataset into separate subsets based on those clusters. Each subset contains the data associated with a particular cluster. This step is crucial for isolating the specific behaviors present in each homogeneous group.
- **Step 3: Apply the Apriori algorithm and FP-Growth to each subset**: With our subsets in place, we apply the Apriori and FP-Growth algorithms to each of them [8]. The goal is to extract meaning association rules within each group. These algorithms analyze transactions in each subset, identifying specific buying or behavior trends for each cluster [10].

Algorithm 2. K-means Algorithm

Input:

- k: The number of clusters to form.
- Data: A set of n data points, each with d dimensions (features).

Output:

- A set of k clusters, where each data point belongs to one of these clusters.
- Centroids: The center of each cluster, serving as a representative point for the cluster.

1. **Initialize Centroids**
   (a) Select k initial centroids. Multiple strategies are available:
      - **Random Initialization**: Randomly select k points from the dataset as centroids.
      - **K-means++ Initialization**: Select an initial point randomly, then choose subsequent points with a preference for being far from existing centroids to improve convergence.

   (b) These centroids are placeholders and will be updated iteratively until convergence.
2. **Assign Points to the Nearest Centroid**
   (a) For each data point in the dataset, calculate the distance to each centroid (commonly using Euclidean distance).
   (b) Assign each data point to the nearest centroid, grouping it into the respective cluster.
3. **Update Centroids**
   (a) After assigning points to clusters, recalculate each centroid by finding the mean position of all points within that cluster. For cluster C_j, the new centroid is:

$$\mu_j = \frac{1}{|C_j|} \sum_{x \in C_j} x \tag{3}$$

   where $|C_j|$ is the number of points in cluster C_j.
   (b) Replace the old centroid with this newly calculated centroid.
4. **Check for Convergence**
   (a) Check if convergence has been reached by:
      - **Centroid Stability**: If the centroids do not change (or change minimally) between iterations, the algorithm can stop.
      - **Maximum Iterations**: Stop if a specified number of iterations has been reached, even if centroids have not stabilized.

   (b) If centroids have changed, return to Step 2 and reassign data points to the nearest centroids. Repeat the assignment and updating steps until convergence criteria are met.

Algorithm 3. Apriori Algorithm

Input: Database of transactions, minimum support (minsup), minimum confidence (minconf)
Output: Frequent itemsets, association rules

1. **Frequent Itemset Generation**
   (a) Generate frequent 1-itemsets by scanning the database.
   (b) $k \leftarrow 2$.
   (c) $C_k \leftarrow$ Candidate item sets of size k.
   (d) $L_{k-1} \leftarrow$ Frequent item sets of size $k-1$.
   (e) **While** L_{k-1} is not empty **do**
      (i) $C_k \leftarrow$ Generate candidates from L_{k-1}.
      (ii) $C_k \leftarrow$ Prune candidates with infrequent subsets.
      (iii) Count the support of each candidate in C_k by scanning the database.
   (f) **Repeat Until No More Candidates:** Continue generating and pruning candidate itemsets until L_k is empty (no frequent itemsets of size k are found).
2. **Association Rule Generation from Frequent Itemsets**
   (a) **Generate Rules from Each Frequent Itemset:** For each frequent itemset L (with size $k \geq 2$), generate all possible non-empty subsets A of L. For each subset A, form the rule $A \rightarrow (L - A)$.
   (b) **Calculate Confidence for Each Rule:** Calculate the confidence for each rule as

$$\text{Confidence}(A \rightarrow (L - A)) = \frac{\text{Support}(L)}{\text{Support}(A)}$$

   If the confidence of the rule meets the minimum confidence threshold, the rule is retained; otherwise, it is discarded.
   (c) **Pruning Rules:** To optimize, only generate rules from subsets of L that are themselves frequent. This reduces unnecessary calculations and ensures that the rules meet the Apriori property.
   (d) **Output:** The result is a set of strong association rules that meet both the minimum support and minimum confidence thresholds.

- **Step 4: Grouping the Rules of Association**: Once we have generated association rules for each subset, we aggregate all these rules to obtain a consolidated view of the entire set of behaviors extracted from each cluster. This consolidation allows us to observe overall trends and correlations among different user groups [4].
- Finally, it is essential to note that our study was carried out using the CC GENERAL dataset, which plays a crucial role in our research because it contains customer data related to credit card transactions. We used this dataset to analyze customer behavior, identify trends in spending patterns, and predict credit risk. It includes credit card data, customers, their balances, purchase and cash advance frequencies, and payment histories [15]. Table 1 shows the columns, and their descriptions of the dataset.

Table 1. Description of the dataset

Attribute	Description
CUST_ID	Unique identifier for each customer
BALANCE	Current balance of the customer's account
BALANCE_FREQUENCY	Frequency of balance updates (e.g., daily, weekly, monthly)
PURCHASES	Total purchases made by the customer
ONEOFF_PURCHASES	Total one-off purchases made by the customer
INSTALLMENTS_PURCHASES	Total purchases made through installment plans
CASH_ADVANCE	Total cash advances taken by the customer
PURCHASES_FREQUENCY	Frequency of purchases (e.g., daily, weekly, monthly)
ONEOFF_PURCHASES_FREQUENCY	Frequency of one-off purchases
PURCHASES_INSTALLMENTS_FREQUENCY	Frequency of purchases made through installment plans
CASH_ADVANCE_FREQUENCY	Frequency of cash advances
CASH_ADVANCE_TRX	Number of cash advance transactions
PURCHASES_TRX	Number of purchase transactions
PAYMENTS	Total payments made by the customer
MINIMUM_PAYMENTS	Minimum payment amount required by the customer
PRC_FULL_PAYMENT	Percentage of customers who made full payments
TENURE	Tenure of the customer's account (e.g., years, months)
CREDIT_LIMIT	Credit limit of the customer's account

Description of each attribute in the dataset.

4 Results and Analysis

In this section, we closely examine the performance of the Apriori algorithm in a distributed environment based on the number of clusters generated by our K-means approach. The goal is to understand how task distribution affects the execution time of the Apriori algorithm when applied to different subsets of data.

- Run on one Cluster: When applying the algorithm to a single cluster, we observe the initial execution time, illustrated in Fig. 2. This configuration serves as a reference point for our comparison.
- Run on two Clusters: By increasing the number of clusters to two, the Fig. 2 illustrates how the distribution of tasks impacts the execution time of the algorithm. We note any significant variation from running on a single cluster.
- Run on three Clusters: Extending our analysis to three clusters, we observe the effect of the increasing complexity of the data distribution on the execution time of the algorithm (Fig. 2). This step determines whether the task distribution continues to improve or whether inefficiencies appear with a finer distribution.

The graph depicts the total execution time of association rule extraction using the Apriori and FP-Growth algorithms as a function of the number of iterations. A priori is faster than FP-Growth for all iterations. The execution time of FP-Growth is approximately 2 times longer than that of Apriori. The execution time for both algorithms increases with the number of iterations.

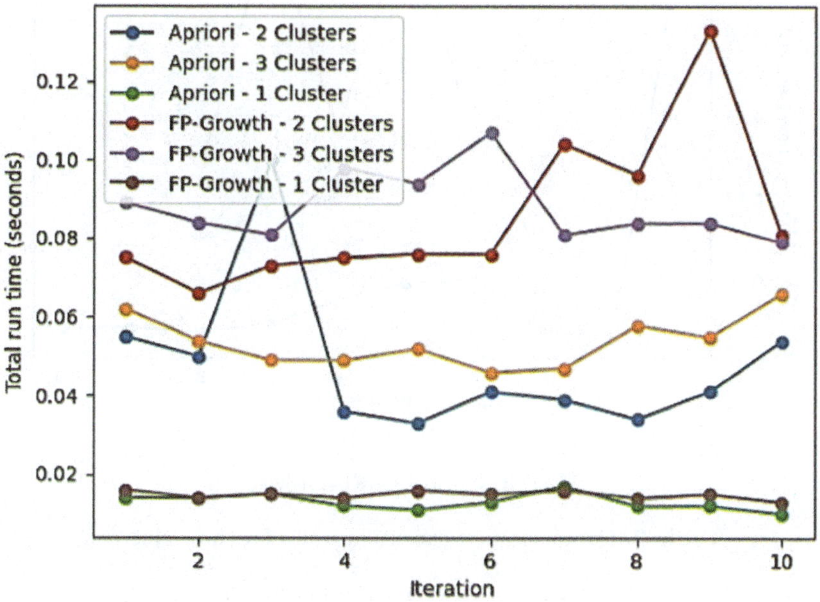

Fig. 2. Comparison of the Apriori Algorithm and the FP-Growth Run-time by the Number of Clusters.

This is because there are more candidates to evaluate at each iteration. The difference in execution time between the two algorithms is more pronounced for later iterations. This is because Apriori is more efficient at handling large candidate sets. A priori is a more efficient algorithm than FP-growth for association rule extraction. The performance gap between the two algorithms is more significant for large datasets. The Fig. 3 shows only the total execution time. It does not illustrate the time spent in different phases of the algorithm, such as candidate generation and pruning.

The algorithm performance may vary depending on the size and nature of the dataset. Table 2 shows a comparison of execution times in seconds for the Apriori and FP-Growth algorithms on three different cluster configurations. Cluster counts are listed in the first column, with corresponding runtimes for Apriori and FP-Growth in subsequent columns. For each number of clusters, the execution time of FP-Growth is generally greater than that of Apriori, although the difference is more pronounced for two clusters.

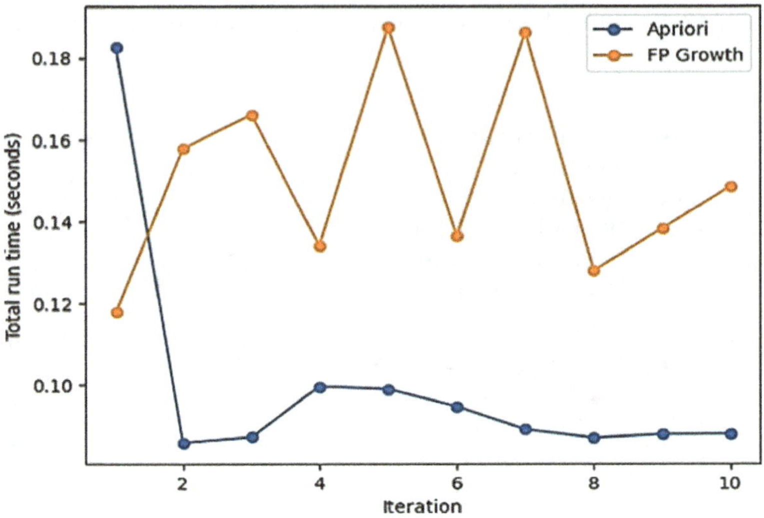

Fig. 3. Apriori vs FP-Growth

Table 2. Comparison of execution times for three clusters

Number of Clusters	Apriori	FP-Growth
1	0.483	0.855
2	0.538	0.881
3	0.130	0.148

To generate a general table of association rules with strong correlation from the dataset, extraction rules are chosen where the confidence and lift values are high, as these metrics indicate a strong association between items. A higher confidence means that the antecedent (the "if" part of the rule) strongly leads to the consequent (the "then" part). A higher lift indicates that the occurrence of the antecedent increases the likelihood of the consequent occurring, beyond what would be expected by chance. Table 3 shows some strong association rules

Table 3. Part of strong association rules

Rule	Antecedents	Consequents	Support	Confidence	Lift
1	{Purchase}	{Payment}	0.45	0.80	1.9
2	{Income}	{Credit_Limit}	0.42	0.78	1.8
3	{Balance}	{Payment}	0.41	0.75	1.7
4	{Credit_Limit}	{Balance}	0.38	0.72	1.6
5	{Purchase, Balance}	{Payment}	0.44	0.74	1.8

using FP-growth by applying filters on confidence (greater than 0.7) and lift (greater than 1.5) with $k = 5$ clusters.

- **Rule 1**: {Purchase} → {Payment}: Customers who make a purchase are highly likely to make a payment. The high confidence (0.80) and lift (1.9) show a strong correlation between these two actions.
- **Rule 2**: {Income} → {Credit_Limit}: Higher income customers tend to have higher credit limits. This rule shows a strong association with a confidence of 0.78 and lift of 1.8.
- **Rule 3**: {Balance} → {Payment}: Customers with a balance are more likely to make payments, with a strong correlation.
- **Rule 4**: {Credit_Limit} → {Balance}: Customers with higher credit limits are likely to have a balance.
- **Rule 5**: {Purchase, Balance} → {Payment}: Customers who make purchases and have a balance are highly likely to make a payment.

5 Conclusion

The preliminary stages of our project were marked by the application of K-means clustering as an essential phase to group datasets exhibiting similar behaviors. This clustering process created homogeneous clusters, laying the groundwork for a more in-depth exploration of association rules. We adopted a systematic approach by combining K-means clustering with powerful association rule extrapolation algorithms, namely, Apriori and FP-Growth. These were chosen for their distinct capabilities: Apriori's strategy for generating and pruning candidate sets and FP-Growth's efficient tree construction.

The choice between Apriori and FP-Growth was guided by considerations specific to our dataset, highlighting the importance of understanding the unique characteristics of each algorithm and adapting accordingly. While Apriori follows an incremental approach to generating and pruning, FP-Growth employs recursive exploration with FP tree construction. Our integrated approach facilitated a thorough exploration of specific behaviors within each cluster via the application of Apriori to datasets already grouped by K-means. This synergy between clustering and association rule extraction strengthened the relevance of the obtained results. The project focused on refining association rule extraction techniques, especially in the context of exploring no distributed datasets. Although we did not introduce entirely new methods, our improvements make a significant contribution to addressing the challenges posed by large-scale datasets. Our study underscores the crucial importance of effectively leveraging established algorithms, paving the way for more efficient data exploration practices.

References

1. Ibrahim Hayatu, H., Mohammed, A., Barroon Isma'eel, A.: Big data clustering techniques: recent advances and survey. In: Machine Learning and Data Mining for Emerging Trend in Cyber Dynamics: Theories and Applications, pp. 57–79 (2021)
2. Jain, A.K., Topchy, A., Law, M.H., Buhmann, J.M.: Landscape of clustering algorithms. In: Proceedings of the 17th International Conference on Pattern Recognition, 2004. ICPR 2004, vol. 1, pp. 260–263. IEEE (2004)
3. Lange, T., Roth, V., Braun, M.L., Buhmann, J.M.: Stability-based validation of clustering solutions. Neural Comput. **16**(6), 1299–1323 (2004)
4. Li, F., Qin, J., Kang, Y.: Closed-loop hierarchical operation for optimal unit commitment and dispatch in microgrids: a hybrid system approach. IEEE Trans. Power Syst. **35**(1), 516–526 (2019)
5. Yin, H., Aryani, A., Petrie, S., Nambissan, A., Astudillo, A., Cao, S.: A rapid review of clustering algorithms. arXiv preprint arXiv:2401.07389 (2024)
6. Hansen, M.H., Yu, B.: Model selection and the principle of minimum description length. J. Am. Stat. Assoc. **96**(454), 746–774 (2001)
7. Tabachnick, B.G., Fidell, L.S., Ullman, J.B.: Using Multivariate Statistics, vol. 6, pp. 497–516. Pearson Boston, MA (2013)
8. Hu, H., Liu, J., Zhang, X., Fang, M.: An effective and adaptable k-means algorithm for big data cluster analysis. Pattern Recogn. **139**, 109404 (2023)
9. Sharma, K., Saini, S., Sharma, S., Kang, H.S., Bouye, M., Krah, D.: Big data analytics model for distributed document using hybrid optimization with k-means clustering. Wireless Commun. Mobile Comput. **2022**(1), 5807690 (2022)
10. Guha, S., Rastogi, R., Shim, K.: ROCK: a robust clustering algorithm for categorical attributes. Inf. Syst. **25**(5), 345–366 (2000)
11. Zhang, T., Ramakrishnan, R., Livny, M.: BIRCH: an efficient data clustering method for very large databases. ACM SIGMOD Rec. **25**(2), 103–114 (1996)
12. Ester, M., Kriegel, H.-P., Sander, J., Xu, X., et al.: A density-based algorithm for discovering clusters in large spatial databases with noise. In: KDD, vol. 96, pp. 226–231 (1996)
13. Pei, J., Han, J., Mao, R., et al.: CLOSET: an efficient algorithm for mining frequent closed itemsets. In: ACM SIGMOD Workshop on Research Issues in Data Mining and Knowledge Discovery, vol. 4, pp. 21–30 (2000)
14. Fakir, Y., Khalil, S., Fakir, M.: Extraction of association rules in a diabetic dataset using parallel FP-growth algorithm under Apache spark. Int. J. Inf. Commun. Technol. (IJICT) **13**(3), 445–452 (2024)
15. https://www.kaggle.com/code/nataliaole/cc-general-clustering

Author Index

A. Laghrib and A. Ghazdali (Eds.): ICMDS 2024, LNNS 1466, pp. 249–250, 2025.
https://doi.org/10.1007/978-3-031-94802-2

The manufacturer's authorised representative in the EU is Springer Nature Customer Service Centre GmbH, Europaplatz 3, 69115 Heidelberg, Germany. If you have any concerns regarding our products, please contact ProductSafety@springernature.com

Printed and bound by CPI Group (UK) Ltd, Croydon, CR0 4YY

30/09/2026

02225573-0003